SATHER CLASSICAL LECTURES
Volume Thirty-two

# The Athenian Year

# THE
# ATHENIAN
# YEAR

Benjamin D. Meritt

UNIVERSITY OF CALIFORNIA PRESS

Berkeley and Los Angeles · 1961

University of California Press
Berkeley and Los Angeles
California

Cambridge University Press
London, England

Printed in the United States of America

# PREFACE

The substance of this book, in somewhat different form, was given in the series of Sather Lectures in Classical Literature at the University of California in March and April of 1959. My warm thanks are due to the Administration of the University and to colleagues at Berkeley for the privilege of spending the spring semester on the Berkeley Campus and for their unfailing helpfulness and hospitality. One cannot have this close association without feeling the inspirational pulse of a great educational center, and without, I think, the humble realization that fortune has been kind to allow participation, in some small measure, in one of the great traditions of American classical scholarship.

My especial thanks must be expressed to the Chairman of the Department of Classics, Professor Arthur E. Gordon, and to Professors L. A. MacKay, W. Kendrick Pritchett, and W. Gerson Rabinowitz. What I owe to Pritchett, in agreement and in disagreement, every page of this book will testify. He and Professor Otto Neugebauer, of Brown University, in their recent book on *The Calendars of Athens*, have solved the thorny problem of dates at Athens κατ᾽ ἄρχοντα and κατὰ θεόν. If I have not shared their other views, I confess my gratitude none the less for this, and always for their clear and forthright presentation of pertinent calendrical evidence.

I acknowledge especially my indebtedness to Neugebauer, with whom I have discussed many of the astronomical questions raised by the study of the Athenian calendar, and to Professor A. E. Raubitschek, of Princeton University, whose help has been invaluable with the various scholia and their interpretation. Miss Margaret Thompson, of the American Numismatic Society, has most graciously put at my disposal the results of her study,

v

now in progress, on Athenian New Style Coinage. And in Greece I have had the unstinted help and coöperation of Markellos Mitsos, Director of the Epigraphical Museum, and of Christos Karouzos, Director of the National Archaeological Museum. There has been no epigraphical text in Athens that I needed to see and study that I did not see and study under their kind dispensation, while the texts from the Agora, of course, have always been readily accessible. Some of the epigraphical work, especially that with the new texts in Chapter IX, was done while more general preparations were in progress for the publication of discoveries in the Agora.

Recent studies have left the available tables of Athenian archons in some confusion. The end, I fear, is not yet. But I think it a useful service to present once more a thoroughgoing table from the fourth to the first century. This appears in Chapter XI, and is available as a new basis for the additions and alterations which new evidence (or further study) may supply. The rest of the volume will speak for itself. My claim for it is not that it everywhere says the last word, but rather that it makes some progress, and presents some new evidence and some old evidence in a new light, in a controversial field.

BENJAMIN D. MERITT

Berkeley, California, May, 1959

# CONTENTS

# ILLUSTRATIONS

# ABBREVIATIONS

A. J. P. = American Journal of Philology

Abh. Ak. Berlin = Abhandlungen der preussischen Akademie der Wissenschaften, philosophisch-historische Klasse

Amer. Hist. Rev. = American Historical Review

Ἀρχ. Δελτίον = Ἀρχαιολογικὸν Δελτίον

Ath. Mitt. = Mitteilungen des deutschen archäologischen Instituts. Athenische Abteilung

Athenaeum = Athenaeum, Studi Periodici di Letteratura e Storia dell' Antiquità

B. C. H. = Bulletin de Correspondance Hellénique

B. S. A. = The Annual of the British School at Athens

Beloch, Gr. Gesch. = Karl Julius Beloch, Griechische Geschichte (second edition, 4 vols., Strassburg, Berlin and Leipzig, 1912-1927)

Busolt-Swoboda, Gr. Staatskunde = Griechische Staatskunde von Georg Busolt, dritte neugestaltete Auflage der griechischen Staats- und Rechtsaltertümer, zweite Hälfte, – bearbeitet von Dr. Heinrich Swoboda (Munich, 1926)

C. I. G. = Corpus Inscriptionum Graecarum

Cl. Phil. = Classical Philology

Cl. Quart. = Classical Quarterly

Cl. Rev. = Classical Review

Corpus – The definitive epigraphical publications of the Berlin Academy are compendiously designated by this title

Dinsmoor, Archons = William Bell Dinsmoor, The Archons of Athens in the Hellenistic Age (Cambridge, Mass., 1931)

Dinsmoor, Athenian Archon List = William Bell Dinsmoor, The Athenian Archon List in the Light of Recent Discoveries (New York City, 1939)

E. M. = Epigraphical Museum at Athens (Inventory)

Frag. gr. Hist. = Die Fragmente der griechischen Historiker

H. S. C. P. = Harvard Studies in Classical Philology

Hesperia = Hesperia, Journal of the American School of Classical Studies at Athens

I. G. = Inscriptiones Graecae

Inscr. Délos = Inscriptions de Délos

J. H. S. = The Journal of Hellenic Studies

Meritt, Athenian Calendar = Benjamin Dean Meritt, The Athenian Calendar in the Fifth Century (Cambridge, Mass., 1928)

Meritt, Athenian Financial Documents = Benjamin Dean Meritt, Athenian Financial Documents of the Fifth Century (Ann Arbor, Michigan, 1932)

Mommsen, Chronologie = August Mommsen, Chronologie, Untersuchungen über das Kalenderwesen der Griechen, insonderheit der Athener (Leipzig, 1883)

P. A. = Prosopographia Attica

Parker and Dubberstein, Babylonian Chronology = Richard Anthony Parker and Waldo Herman Dubberstein, Babylonian Chronology 626 B.C.–A.D. 75 (Providence, Rhode Island, 1956)

Pauly-Wissowa, R. E. = Paulys Real-Encyclopädie der classischen Altertumswissenschaft, Neue Bearbeitung (Stuttgart, from 1894), edited by Georg Wissowa and others

Polemon = Πολέμων, 'Αρχαιολογικὸν Περιοδικόν

Pritchett and Meritt, Chronology = William Kendrick Pritchett and Benjamin Dean Meritt, The Chronology of Hellenistic Athens (Cambridge, Mass., 1940)

Pritchett and Neugebauer, Calendars = W. Kendrick Pritchett and O. Neugebauer, The Calendars of Athens (Cambridge, Mass., 1947)

R. E. G. = Revue des Études Grecques

Rev. crit. = Revue critique d'histoire et de littérature

Rh. Mus. = Rheinisches Museum für Philologie

S. E. G. = Supplementum Epigraphicum Graecum

S. V. F. = Stoicorum Veterum Fragmenta

Sitz. Ak. Berlin = Sitzungsberichte der preussischen Akademie der Wissenschaften, philosophisch-historische Klasse

# CHAPTER I

## The Reckoning of Time

There were at Athens, beginning at least as early as the fifth century before Christ, two principal ways of dating public events. One must fix them well in mind, for however much our argument may seem to digress or to get lost in details these two systems of reckoning time run through the whole fabric of Athenian history. Their separate development and their interrelation form much of the substance of this and subsequent chapters.

First, time was reckoned by months. These months were governed by the moon and corresponded roughly to the intervals between first visibilities of the new lunar crescent. There were twelve such months in a normal or ordinary year, and, since an average lunar cycle is just slightly more than 29½ days, a year of twelve such months contained normally 354 days, half of the months of 30 days and half of 29. We need not yet go into the question of ordinary years—this has become the accepted technical terminology—of 353 days, in which I do not believe, or of 355 days, in which I do believe, for these will become part of the argument later. One could speak of the year as a lunar year, except that this terminology has implications that are too narrowly astronomical. One may call it the religious year, or the festival year, for the numerous festivals of Athens were dated in terms of its calendar. In order to avoid confusion I have used the term "festival year" or "festival calendar" when referring to the year of months which ran from Hekatombaion (the first month) to Skirophorion (the last month).

Since this year of 354 days fell short by some 11¼ days of parity with the seasonal year, which depended upon the sun, it was necessary from time to time to have a festival year of 13 months to keep events in the festival year in time with the seasons. These intercalary years, so called because of the intercalation of an extra (thirteenth) month, normally contained 384 days, and in the course of time the astronomers set up ideal cycles specifying the sequence of such intercalations. Meton's cycle of nineteen years, propounded at Athens with its initial date in 432 B.C., is a case in point.[1] Beginning with Skirophorion 13 in 432 B.C., seventeen of these cycles had elapsed by Skirophorion 14 of 109 B.C., the measurement being from summer solstice to summer solstice for the seasonal years. This span is covered by an astronomical almanac, inscribed on stone at Miletos, the so-called Milesian *parapegma*, which reads as follows:[2] " [from] the summer solstice in the archonship of Apseudes on Skirophorion 13, which according to the Egyptians was Phamenoth 21, till that in the archonship of Polykleitos on Skirophorion 14, Pauni 11 according to the Egyptians – – – –." Meton's cycle called for a regular and predictable succession of days, months, and years. It can be demonstrated from the epigraphical records that the Athenians did not in practice arrange their days, months, and years in a precisely regular or predictable pattern. Apparently they enjoyed liberty to add or not to add the intercalary year—this too has become the accepted technical terminology— whenever and wherever they pleased.[3] This freedom from

[1] See Diodoros, XII, 36, 2: ἐν δὲ ταῖς Ἀθήναις Μέτων ὁ Παυσανίου μὲν υἱός, δεδοξασμένος δὲ ἐν ἀστρολογίᾳ, ἐξέθηκε τὴν ὀνομαζομένην ἐννεακαιδεκαετηρίδα, τὴν ἀρχὴν ποιησάμενος ἀπὸ μηνὸς ἐν Ἀθήναις Σκιροφοριῶνος τρισκαιδεκάτης.

[2] [θ]ερινῆς τρο[π]ῆς [γε]νομένης ἐπὶ Ἀψεύδους Σκιροφοριῶνος ΙΓ, ἥτις ἦν κατὰ τοὺς Αἰγυπτίους μία καὶ Κ̄ [τ]οῦ Φαμενώθ, ἕως [τῆ]ς γενομένης ἐπὶ [Πολ]υκλείτου Σκι[ροφορι]ῶνος ΙΔ, κα[τὰ δὲ τοὺ]ς Αἰγυπτί[ους τοῦ Παυ]νὶ τῆς ΙΑ – – – –. See W. B. Dinsmoor, *Archons of Athens*, p. 226 note 2, for the error in the text of Meritt's *Athenian Calendar*, p. 88. The solstices were of long standing fixed points in defining the seasonal year. Hesiod knew them, and called them τροπαὶ ἠελίοιο. He also assumed that his hearers knew them. And a device for timing the solstice is mentioned in the Odyssey (cf. H. T. Wade-Gery, *Essays in Greek History* [Oxford, 1958], pp. 1-3).

[3] B. D. Meritt, *The Athenian Calendar* (1928), p. 123: "attempts to build

cyclical restriction has been confirmed by the studies of Dinsmoor [4] and Pritchett and Neugebauer,[5] who add (i. e., Pritchett
and Neugebauer) that the Athenians did not attempt a systematic
control of their festival calendar even at the beginnings and
ends of cycles. I am in complete agreement with them. Indeed,
I held in 1928 that the very first Metonic cycle had eight rather
than the theoretically correct seven intercalary years—an opinion
which I still hold—and there is evidence, I believe, now, that
one of the theoretical cycles in the second century must also
be restored with eight intercalary years.[6] These vagaries have
to be compensated by cycles in which there were only six intercalary years. For the workaday calendar at Athens, the calendar
of our literary texts and of the inscriptions, this means that the
Metonic cycles have no value as evidence for the calendar
character of any individual year.

Normally, when a year had thirteen months it was the sixth
month, Posideon, which was repeated, called Ποσιδεὼν ὕστερος,
or Ποσιδεὼν ἐμβόλιμος, or Ποσιδεὼν δεύτερος. But even this was
not a hard and fast rule. Other months, as well as Posideon,
could be chosen for doubling, and sometimes we are able to
discern the reasons for the irregular choice.

Secondly, time was reckoned by the political year, the year
of the Council. From about the end of the fifth century this
year was coterminous with the festival year. It may be defined
as the conciliar year—and we may count this definition too as
part of our technical terminology. The divisions of the year
were called prytanies, and when there were ten phylai, or
political subdivisions of the Athenian citizen body (Erechtheis
to Antiochis), represented in the Council, as was the case from

up any calendar scheme merely by means of cycles of intercalation are destined
to failure."
[4] *Archons of Athens*, p. 320.
[5] *The Calendars of Athens*, pp. 7-10.
[6] See below, p. 182, and the table on pp. 235-236 for the years 185/4-167/6.

the beginning of our investigation down to the creation of the so-called Macedonian phylai in 307 B.C., the year was divided into ten periods of time, as nearly equal as might be, and these were distributed by lot during the course of the year, one so-called prytany to each phyle.

When there were twelve phylai, after the creation of Antigonis and Demetrias, then the prytanies in ordinary years were approximated closely to the months, and our problems would have been simple if they had in fact been identical. Sometimes they were and sometimes they were not, and the correspondences and divergencies become part of our investigation. Then there was a period of thirteen phylai after the creation of Ptolemais in 223 B.C., followed by a minimum of time in 201 B.C. when there were only eleven phylai and an epoch from then on to the end of the Hellenistic Age (which is the limit of this investigation) when there were twelve phylai again after the creation of Attalis.

Thus the festival year, which was divided into twelve or thirteen months, and which had 354 (355) or 384 days, was equated with the conciliar year which was divided in a different way into prytanies, each prytany being given the 10th, 12th, 13th, or 11th part of 354 (355) or 384 days, as the case might be. These relationships held valid for those centuries when the festival year and the conciliar year were coterminous. In the fifth century the conciliar year was different from the festival year, and its relation to the festival year requires special study.

It will be well to take a sample of dating by month and prytany so that the interrelation of these two systems may be made clear.

A decree in honor of some otherwise unknown man has been in part preserved on stone from the year 326/5 B.C. Its opening lines read as follows: [7]

---

[7] For the text in line 6 see Pritchett and Neugebauer, *Calendars*, p. 54.

## *I.G.*, II², 359

*a.* 326/5 *a.* ΣΤΟΙΧ. 20

['Α] σ τ υ [– – – – – –]
'Επὶ Χρέμητο[ς ἄρχοντος ἐ]
πὶ τῆς 'Ερεχθη[ίδος ἑβδόμ]
ης πρυτανείας· ['Ελαφηβολ]
5  [ι]ῶνος ὀγδόηι ἱ[σταμένου,]
[τ]ριακοστῆι τῆ[ς πρυτανέ]
[ας·] ἐκκλησία κυ[ρία· τῶν πρ]
[οέδρω]ν ἐπεψή[φιζεν – – – –]

The month is the ninth (Elaphebolion) and the prytany is the seventh (Erechtheis). In any given year the order of the prytanies among the phylai depends on allotment;[8] we know that the month was the ninth and the prytany the seventh because these restorations are unique for the equation given in lines 4-7. We are now in that era of the ten phylai when the festival year and the conciliar year began on the same day (Hekatombaion 1 = Prytany I 1) and when the prytanies each counted a tenth of the year (36 or 35 days), the total of the ten prytanies amounting normally to 354 days.

If the first four prytanies had each 36 days and the rest each 35 days then the 30th day of the seventh prytany was the 244th day of the year. If the months alternated between 29 and 30 days, then the 8th day of Elaphebolion was likewise the 244th day of the year, and the equation implicit in the epigraphical text is satisfied. There is no problem about the calendar of this year, so far as the evidence goes. It was an ordinary year of twelve months, which alternated between hollow and full (29 and 30 days), and the first four prytanies each had 36 days:

---

[8] See Sterling Dow, *Hesperia*, Suppl. I (1937), pp. 210-211. See below, p. 8 note 9.

| Months | 29 | 30 | 29 | 30 | 29 | 30 | 29 | 30 | 29 | 30 | 29 | 30 = 354 |
| Prytanies | 36 | 36 | 36 | 36 | 35 | 35 | 35 | 35 | 35 | 35 = 354 |

This pattern of the conciliar year, in which the first four prytanies each had 36 days, follows exactly the definition of the conciliar year as given by Aristotle. He affirms [9] that in his day the ten prytanies of the year were so divided that the first four had 36 days each and the last six 35 days each. Surely this was one normal arrangement: an ordinary year of twelve months, with which the conciliar year was coterminous, and both of which could be defined by the name of the archon who held office from Hekatombaion 1 (= Prytany I 1) to Skirophorion ἔνη καὶ νέα (Prytany X 35). Aristotle says nothing about an ordinary year that may have had 355 days, or perhaps as few as 353 days. His rule, therefore, must show some slight variation from actual fact, if and when either of these eventualities came to pass. In a year of 355 days, one might perhaps posit that the fifth prytany had 36 days, like the four preceding, and that the last five each had 35 days. Or, in view of the freedom which the Athenians allowed themselves in other calendar matters, one might posit that the extra prytany of 36 days could have been any one of those from fifth to tenth, perhaps chosen by lot as were the phylai which were to function during those remaining subdivisions of the conciliar year. Pritchett and Neugebauer have assumed "as a working hypothesis" that the adjustment of the conciliar year to an ordinary year of 353 or 355 days was effected by subtracting or adding a day in the tenth, or final, prytany, and they tabulate the days of an ordinary year as follows: [10]

---

[9] Ἀθ. Πολ., 43, 2: βουλὴ δὲ κληροῦται πεντακόσιοι, πεντήκοντα ἀπὸ φυλῆς ἑκάστης. πρυτανεύει δ' ἐν μέρει τῶν φυλῶν ἑκάστη καθ' ὅ τι ἂν λάχωσιν, αἱ μὲν πρῶται τέτταρες ἓξ καὶ τριάκοντα ἡμέρας ἑκάστη, αἱ δὲ ἓξ αἱ ὕστεραι πέντε καὶ τριάκοντα ἡμέρας ἑκάστη· κατὰ σελήνην γὰρ ἄγουσιν τὸν ἐνιαυτόν.

[10] Pritchett and Neugebauer, Calendars, p. 36.

| Number of Prytany | Length of Prytany | Number of First Day of Prytany | Number of Last Day of Prytany |
|---|---|---|---|
| I | 36 | 1 | 36 |
| II | 36 | 37 | 72 |
| III | 36 | 73 | 108 |
| IV | 36 | 109 | 144 |
| V | 35 | 145 | 179 |
| VI | 35 | 180 | 214 |
| VII | 35 | 215 | 249 |
| VIII | 35 | 250 | 284 |
| IX | 35 | 285 | 319 |
| X | $35 \pm 1$ | 320 | $354 \pm 1$ |

I believe that, even as a working hypothesis, this is too rigid a scheme. Wherever the extra day was added in the longer year, I would merely ask that at the end of the year the last day of the final month should fall on the last day of the final prytany. Nor does Aristotle say anything about the intercalary year, which normally had 384 days. If he was giving a general rule, this omission is perhaps less remarkable. One could be expected to infer, by analogy, that the intercalary year would have four prytanies of 39 days, followed by six prytanies of 38 days. Pritchett and Neugebauer call this a reasonable assumption, as indeed I too believe that it is. They then make allowance for divergence in case of intercalary years of 383 and 385 days, adding or subtracting a day in the tenth, or final, prytany, and they present the following table to show the days of an intercalary year: [11]

| Number of Prytany | Length of Prytany | Number of First Day of Prytany | Number of Last Day of Prytany |
|---|---|---|---|
| I | 39 | 1 | 39 |
| II | 39 | 40 | 78 |
| III | 39 | 79 | 117 |
| IV | 39 | 118 | 156 |
| V | 38 | 157 | 194 |
| VI | 38 | 195 | 232 |
| VII | 38 | 233 | 270 |
| VIII | 38 | 271 | 308 |
| IX | 38 | 309 | 346 |
| X | $38 \pm 1$ | 347 | $384 \pm 1$ |

[11] Pritchett and Neugebauer, *Calendars*, p. 37.

As in the case of the ordinary year, I again express skepticism. If there was to be an extra day, I believe that the Athenians might have added it to any one of the last six prytanies. The evidence advanced for subtracting a day from the final prytany to match a festival year of 383 days seems to me not conclusive. It is found in the intercalary years 341/0 and 336/5, where the last day of the last month was equated with the 37th day of the last prytany, a circumstance which seems to be in agreement with the hypothesis of an intercalary year of 383 days.

The evidence for 341/0 is in the preamble of the inscription now published as *I.G.*, II², 229, with *Addenda* (p. 659):

*I.G.*, II², 229, with *Addenda* (p. 659)

a. 341/0 a.                                          ΣΤΟΙΧ. 38

[Ἐπὶ Νικομάχου ἄρχοντος ἐπ]ὶ τῆς Λεων[τίδ]ο[ς δεκ]
[άτης πρυτανείας, ἧι Ὀνήσι]ππος Σμικύ[θο] Ἀ[ραφήν]
[ιος ἐγραμμάτευεν· ἕνηι κα]ὶ [ν]έαι, ἐβδ[όμηι] κ[αὶ τρ]
[ιακοστῆι τῆς πρυτανεία]ς· – – – – – κτλ. – – – – – –

The name of the month was omitted, but was almost certainly Skirophorion. No good calendar equation can be formulated if the prytany is restored as the fifth (πέμπτης), and a restoration as seventh (ἑβδόμης) is excluded because in this year it is known that Pandionis held the seventh prytany (*I.G.*, II², 228). The date ἔνη καὶ νέα may indeed have been the last day of the month; but it may also have been the next to the last day, followed by the true last day ἔνη καὶ νέα ἐμβόλιμος. Such an assumption would permit a normal intercalary year of 384 days, corresponding accurately to the (hypothetical) Aristotelian definition of four prytanies of 39 days followed by six prytanies of 38 days.

Even more instructive is the evidence for 336/5. There are three equations for this year, restored and interpreted most recently as follows: [12]

---

[12] Pritchett and Neugebauer, *Calendars*, pp. 43-44.

(1)

*I.G.*, II², 328

*a.* 336/5 *a.* ΣΤΟΙΧ. 28

['Επὶ Πυθοδήλου ἄρχοντος ἐπ]ὶ τῆς 'Α[κ]
[αμαντίδος τετάρτης πρυτ]ανείας, ἧ
[ι ................¹⁹......... ἐ]γραμμάτ
[ευεν· Μαιμακτηριῶνος τετ]ράδι φθί
5 [νοντος, ὀγδόει καὶ εἰκοστ]εῖ τῆς πρ
[υτανείας· – – – – – – – –] – κτλ. –

[Maimakterion] 27 = Prytany [IV 28] = 145th day

The restorations are almost certainly correct, and a great improvement over those previously given.

(2)

*I.G.*, II², 330, lines 29-30

*a.* 336/5 *a.* ΣΤΟΙΧ. 48

['Επ]ὶ Πυθοδήλου ἄρχοντος [ἐπὶ τῆς ... ηίδος ἐνάτης πρυτανεί]
[ας,] τετράδι ἐπὶ δέκα, δευτ[έραι τῆς πρυτανείας· – – – – –]

⟨Mounichion⟩ 14 = Prytany [IX] 2 = 310th day

(3)

*I.G.*, II², 330, lines 47-49

*a.* 336/5 *a.* ΣΤΟΙΧ. 48

['Επ]ὶ Πυθοδήλου ἄρχοντος ἐπ[ὶ τῆς ...⁶... ίδος δεκάτης πρυτα]
[ν]είας, ἕνει καὶ νέαι, ἑβδόμη[ι καὶ τριακοστεῖ τῆς πρυτανεία]
[ς·] – – – – – – – – – – – – κτλ. – – – – – – – – – – – – – –

⟨Skirophorion⟩ 29 = Prytany [X] [3]7 = 383rd day

To bring into focus some of the doubts which I think may be entertained about recent calendar studies, I wish to propose, tentatively, another interpretation of these three equations. I have no changes to offer in the restorations, which I believe correct.

Let us assume that the first equation should, in fact, read

[Maimakterion] 27 = Prytany [IV 28] = 145th day

while the second equation, as recorded, remains

⟨Mounichion⟩ 14 = Prytany [IX] 2 = 310th day.

Here we accept the hypothetical Aristotelian rule that the first four prytanies each had 39 days and that the next four each had 38 days, so that the days in the year may fall as indicated. Of the six months from Maimakterion to Elaphebolion inclusive four were full and two were hollow, a sequence which, in the latest studies,[13] has been followed by three successive hollow months at the end of the year before the equation

⟨Skirophorion⟩ 29 = Prytany [X] 37 = 383rd day.

The three successive hollow months at the end of the year could be obviated by adding another day both to Skirophorion and to the tenth prytany, just as we suggested as possible (above, p. 10) for the year 341/0. Indeed, this would relieve the festival calendar of inaccuracy, for the last three months of this year 336/5 were, in fact, astronomically, not all hollow, and the year had 384 (not 383) days.

Times of first possible observation of the lunar crescent have been published for Babylon by Parker and Dubberstein, who report (among others) the dates during the year 336/5 B.C.[14] Since the lunar months were essentially the same, in relative lengths throughout the year, at Athens, I give the dates and add the names of the Attic months with their lengths in days:

| July | 6 | Hekatombaion | 30 |
|------|---|--------------|-----|
| August | 5 | Metageitnion | 29 |
| September | 3 | Boedromion | 30 |
| October | 3 | Pyanepsion | 29 |

[13] See Pritchett and Neugebauer, *Calendars*, p. 43.
[14] R. A. Parker and W. H. Dubberstein, *Babylonian Chronology* (Providence, 1956), pp. 35-36.

| November | 1  | Maimakterion | 30 |
|----------|----|----|----|
| December | 1  | Posideon | 29 |
| December | 30 | Posideon II | 30 |
| January  | 29 | Gamelion | 30 |
| February | 28 | Anthesterion | 29 |
| March    | 29 | Elaphebolion | 30 |
| April    | 28 | Mounichion | 29 |
| May      | 27 | Thargelion | 30 |
| June     | 26 | Skirophorion | 29 |

Since I do not hold that the Athenians depended on empirical observation of the crescent I do not think it essential to believe that the Babylonian lengths applied at Athens, but I do hold that they are hard to reconcile in 336/5 if one has as his hypothesis a festival year which depends on the observed crescent and then posits three hollow months in succession [15] as well as faulty observation at either the beginning or end to yield only 383 rather than the correct number of 384 days.[16] This latter fault could be corrected by taking the last equation to represent not the last day, but the next to last day, of the year. But this still leaves the two hollow months of Mounichion and Thargelion juxtaposed, whereas one of them, astronomically, should have been full.

But these observations which attribute the stigma of poor performance, however slight it may seem, to those at Athens who were responsible for observing the new lunar crescent are all based on three presuppositions, which should now again be brought under question:

1. The beginning of each month in Athens was fixed by actual observation of the lunar crescent.

2. The date τετρὰς φθίνοντος was the 27th day of the month, and not 26th, regardless of whether the month was full or hollow.

3. Aristotle's statement that the first four prytanies were of

[15] Pritchett and Neugebauer, *Calendars*, pp. 43-44.
[16] Pritchett and Neugebauer, *Calendars*, p. 43 note 19.

36 days and the last six prytanies of 35 days must be rigidly upheld—no matter what the consequences—even to the extent of insisting on the first four prytanies of 39 days and the last six of 38 days in an intercalary year.

If, for the sake of argument, we deliberately violate all three of these presuppositions, the first calendar equation of the year 336/5 may be formulated thus:

[Maimakterion] 26 = Prytany [IV 28] = 144th day.

Here the year begins with hollow Hekatombaion, and proceeds with regular alternation of full and hollow months through hollow Maimakterion. The fourth day from the end of Maimakterion (29, 28, 27, 26, – – –) is the 26th and is the 144th day of the year. It will follow that the first two prytanies each had 39 days and that the third prytany had 38 days.

The second equation is

⟨Mounichion⟩ 14 = Prytany [IX] 2 = 309th day.

The prytanies from IV to VIII inclusive will four of them have had 38 days each and one will have had 39 days. Hence the second day of the ninth prytany will be reckoned as $39 + 39 + 38 + 4(38) + 39 + 2 = 309$. In the meantime the months proceed with regular alternation so that the 14th day of Mounichion will be reckoned as $29 + 30 + 29 + 30 + 29 + 30 + 29 + 30 + 29 + 30 + 14 = 309$.

The third equation is

⟨Skirophorion⟩ 29 = Prytany [X] [3]7 = 383rd day.

There were 15 days remaining in hollow Mounichion after the 14th, 30 days in full Thargelion, and 29 days in hollow Skirophorion from the time of the second to the time of the third equation ($309 + 15 + 30 + 29 = 383$). If the ninth prytany had 39 days and the tenth prytany 38 days, then the date Prytany X 37 also falls on the 383rd day (309 + 37 + 37 = 383). The equations as preserved indicate a year in which months and prytanies were arranged as follows:

| | | | |
|---|---|---|---|
| 29 | Hekatombaion | Prytany I | 39 |
| 30 | Metageitnion | Prytany II | 39 |
| 29 | Boedromion | | |
| 30 | Pyanepsion | Prytany III | 38 |
| 29 | Maimakterion | Prytany IV | 38 |
| | [Maimakterion] 26 = Prytany [IV 28] | | |
| 30 | Posideon | Prytany V | 38 |
| 29 | Posideon II | | |
| 30 | Gamelion | Prytany VI | 38 |
| 29 | Anthesterion | Prytany VII | 38 |
| 30 | Elaphebolion | | |
| 29 | Mounichion | Prytany VIII | 39 |
| | ⟨Mounichion⟩ 14 = Prytany [IX] 2 | | |
| 30 | Thargelion | Prytany IX | 39 |
| 29 + 1 | Skirophorion | Prytany X | 38 |
| | ⟨Skirophorion⟩ 29 = Prytany [X] [3]7 | | |
| ___ | | | ___ |
| 384 | Total Days | | 384 |

I suggest this as a valid solution to the calendar problems of the year 336/5 B.C., with the corollary that the three current presuppositions which it violates are in error. These will be studied further in the following chapters.

# CHAPTER II

## The First of the Month

If the presupposition be true that the beginning of each month in Athens was fixed by observation of the lunar crescent, then it means that the festival calendar of Athens was the same as the true lunar calendar. This identification will be valid whether we define the day on which each month began as the day of actual conjunction of sun and moon or as the evening when the young crescent was first visible after the conjunction. Meton fixed the epochal date of his astronomical cycle on the 13th day (inclusive) after the conjunction of June 15 in 432 B.C. But Pritchett and Neugebauer stress that the Athenians could not have calculated the complex tables of all dates of true conjunction. On the other hand, the archons could have watched through the waning days of the old moon and then (more or less prepared for the phenomenon) have begun the new month on that evening when the new crescent was first visible. Astronomically, a lunar calendar made up of months thus ended and thus begun might show marked divergences from a regular alternation of months of 29 and 30 days. Pritchett and Neugebauer, in their exposition of the calendar, have chosen at random, from Babylonian tables of 191/0 and 104/3 B.C., sequences of months which have the following numbers of days:

29   30   29   30   30   30   29   30   29   29   30   29   29
and
30   30   29   30   29   30   29   29   30   29   30   29   30

and they have observed that sequences amounting to four consecutive 30-day months or to three consecutive 29-day months are possible, though rare, while pairs of 30-day months or 29-day

months are very common. The Babylonian and the ancient
Jewish calendars are classic examples of this strictly precise
method of determining the beginnings and lengths of months.[1]

In Babylonia and in Israel there is abundant information about
the pains taken to make certain that the observations of the
young crescent were correct.[2] There is no evidence for anything of this sort in Athens. With all her literature, and with
all the archaeological and epigraphical evidence now available,
it is astonishing (were observation the rule) that nothing is said
about observing the lunar crescent to fix the ending or the
beginning of a month. One must conclude, a priori, that the
Athenians did not care for the perpetual and irregular precision
which constant observation could afford, and that they relied
on some other method in determining the lengths of their
months, a method not tied to the intricacies of true lunations,
either calculated or observed.

To show the contrast between the Athenian's apparent mild
indifference and the care that was taken in Israel, for example,
to determine the time of first observation of the crescent moon,
I quote several sentences (a small part only) of the account of
Jewish practice given by Maimonides in his treatise on *Laws
Concerning the Sanctification of the New Moon* (Chapter II): [3]

> " Two worthy men only, qualified to function as wit-
> nesses in any other legal matter, were fit to testify con-
> cerning the new moon. Women and slaves were considered
> disqualified as witnesses, and their testimony could not be
> accepted. If father and son had seen the new moon, they

---

[1] See Pritchett and Neugebauer, *Calendars*, pp. 5-7.
[2] See F. K. Ginzel, *Handbuch der mathematischen und technischen Chronologie* (II, 1911), pp. 40-52.
[3] Translated by Solomon Gandz, with supplementation and an introduction
by Julian Obermann, and an astronomical commentary by Otto Neugebauer
(New Haven, 1956). Maimonides composed his treatise in A.D. 1178, but
chapters I-V " expound the rules concerning the calendar that are prescribed
or implied in Scripture or handed down by Mosaic tradition " (p. xvi).

were to go to court and testify. . . . Originally, . . . the court used to accept evidence concerning the new moon from any Israelite, for the legal presumption was that any Israelite is qualified as a witness until evidence to the contrary is brought to light. However, when heretics began to cause trouble in a mischievous manner and to hire men to testify that they had observed the new crescent, although they did not in fact see it, the Sages decreed that evidence concerning the new moon should not be accepted unless the witnesses were known to the court as worthy men, and that the witnesses should be duly tested and examined.

". . . The court used to employ methods of calculation of the kind employed by astronomers in order to ascertain whether the new moon of the coming month would be seen to the north or to the south of the sun, whether its latitude would be wide or narrow, and in which direction the tips of its horns would point. And when witnesses appeared in order to testify, the court used to examine them as follows: Where did you see the new moon, to the north or to the south? In which direction did the horns point? How great was its altitude, in the estimate of your eyes, and how wide its latitude?

". . . If the witnesses said that they had seen the new moon reflected in the water, or in the clouds, or in a crystal, or if they said that they had seen part of it in the sky and part in the clouds, or in the water, or in a crystal, this was not considered a valid observation."

And much, much more. But of all this there is no trace in Athens. Whatever the Athenian system for fixing the ending and the beginning of a lunar month, it was certainly different from the Jewish, and not so encumbered with ritual.

In the search for positive evidence, an attempt has nevertheless been made recently to show that the Athenians used actual observations of the new crescent to determine the beginnings of their months, the indications being implicit in four calendar equations derived from Ptolemy's *Almagest* for the years from 295 to 283 B.C.:

| New Moon | | | First Day of Month |
|---|---|---|---|
| 295 B.C. | Nov. 26 | 7:45 A.M. | Nov. 27 |
| 294 B.C. | Feb. 23 | 2:30 A.M. | Feb. 24 |
| 283 B.C. | Jan. 22 | 11:15 A.M. | Jan. 23 |
| 283 B.C. | Oct. 14 | 1:30 P.M. | Oct. 16/17 |

Since "the beginning of the month has in each case exactly the position with respect to the new moon which one would normally expect for lunar months" the argument has been that "one should consider these dates as an astronomical proof of the lunar character of the civil calendar."[4] I believe that this presses the evidence too hard. Surely, over the years, even if the Athenians had merely guessed at the time of new moon, they would have been right much of the time. But it is possible to demonstrate that these Ptolemaic dates have, in fact, nothing whatever to do with the festival calendar at Athens. They must be studied again.

Ptolemy did not give the dates of new moon, but in four separate paragraphs he reported observations of the stars that had been made by the astronomer Timocharis at Alexandria:

1. *Almagest*, VII (ed. Heiberg), p. 32: πάλιν Τιμόχαρις μέν φησιν ἐν 'Αλεξανδρείᾳ τηρήσας, ὅτι τῷ λς' ἔτει τῆς πρώτης κατὰ Κάλιππον περιόδου τοῦ μὲν Ποσειδεῶνος τῇ κε', τοῦ δὲ Φαωφὶ τῇ ις', ὥρας ι' ἀρχούσης ἀκριβῶς σφόδρα ἐφαίνετο κατειληφυῖα ἡ σελήνη τῇ βορείῳ ἀψῖδι τὸν πρὸς ἄρκτον τῶν ἐν τῷ μετώπῳ τοῦ Σκορπίου. καί ἐστιν ὁ χρόνος κατὰ τὸ νυδ' ἔτος ἀπὸ Ναβονασσάρου κατ' Αἰγυπτίους Φαωφὶ ις' εἰς τὴν ιζ' μετὰ γ̄ ὥρας καιρικὰς τοῦ μεσονυκτίου – – – .

"Again Timocharis reports an observation that he made at Alexandria, that in the 36th year of the first Kallippic cycle, on

---

[4] Pritchett and Neugebauer, *Calendars*, p. 12; cf. W. B. Dinsmoor, *Archons*, p. 365.

the 25th day of the month Posideon, the 16th day of the month Phaophi, at the beginning of the 10th hour the moon appeared exactly to have reached with its northern rim the northernmost star in the forehead of Scorpio. The time is during the 454th year of the era of Nabonassar, according to the Egyptians from the 16th to the 17th of the month of Phaophi, at three seasonal hours after midnight – – – – ."

I Kall. 36, Pos. 25, Phaophi 16, beginning of the 10th night hour = Nab. 454, Phaophi 16/17 = 295 B.C., Dec. 21, 3:24 A.M.[5]

2. *Almagest*, VII (ed. Heiberg), p. 28: πάλιν Τιμόχαρις μὲν ἀναγράφει τηρήσας ἐν ᾿Αλεξανδρείᾳ, διότι τῷ λϛ᾽ ἔτει τῆς πρώτης κατὰ Κάλιππον περιόδου τοῦ μὲν ᾿Ελαφηβολιῶνος τῇ ιε᾽, τοῦ δὲ Τυβὶ τῇ ε᾽, ὥρας γ᾽ ἀρχομένης ἡ σελήνη μέσῃ τῇ πρὸς ἰσημερινὴν ἀνατολὴν ἀψῖδι τὸν Στάχυν κατέλαβεν, καὶ διῆλθεν ὁ Στάχυς ἀφαιρῶν αὐτῆς τῆς διαμέτρου πρὸς ἄρκτους τὸ τρίτον μέρος ἀκριβῶς. καί ἐστιν ὁ χρόνος κατὰ τὸ υνδ᾽ ἔτος ἀπὸ Ναβονασσάρου κατ᾽ Αἰγυπτίους Τυβὶ ε᾽ εἰς τὴν ϛ᾽ πρὸ δ̄ ὡρῶν καιρικῶν τε καὶ ἰσημερινῶν ἔγγιστα τοῦ μεσονυκτίου – – – – .

"Again Timocharis reports an observation made by him at Alexandria, that in the 36th year of the first Kallippic cycle, on the 15th day of the month Elaphebolion, the 5th day of the month Tybi, at the beginning of the 3rd hour, the moon reached with the middle of its eastern rim the star Spica, and Spica passed behind it cutting off one third part exactly of its diameter toward the north. The time is during the 454th year of the era of Nabonassar, according to the Egyptians from the 5th to the 6th of the month Tybi, four seasonal and equinoctial hours, very nearly, before midnight – – – – ."

---

[5] See W. B. Dinsmoor, *Archons*, p. 365. Since the observation is dated in terms of the Egyptian calendar, the Julian equivalent is readily determinable.

I Kall. 36, Elaph. 15, Tybi 5, beginning of the third night
hour = Nab.454, Tybi 5/6 = 294 B.C., March 9, 8:00 P.M.

3. *Almagest*, VII (ed. Heiberg), p. 25: Τιμόχαρις μὲν γὰρ
ἀναγράφει τηρήσας ἐν ᾿Αλεξανδρείᾳ ταῦτα, διότι τῷ μζ´ ἔτει τῆς
πρώτης κατὰ Κάλιππον ἑξκαιεβδομηκονταετηρίδος τῇ η´ τοῦ ᾿Ανθε-
στηριῶνος, κατ᾿ Αἰγυπτίους τῇ κθ´ τοῦ ᾿Αθύρ, ὥρας γ´ ληγούσης τὸ
νότιον μέρος ἥμισυ τῆς σελήνης ἐπιβεβηκὸς ἐφαίνετο ἐπὶ τὸ ἑπόμενον
ἤτοι γ´ ἢ Ľ μέρος τῆς Πλειάδος ἀκριβῶς. καί ἐστιν ὁ χρόνος κατὰ
τὸ υξε´ ἔτος ἀπὸ Ναβονασσάρου κατ᾿ Αἰγυπτίους ᾿Αθὺρ κθ´ εἰς τὴν
λ´ πρὸ τριῶν ὡρῶν τοῦ μεσονυκτίου καιρικῶν – – – – .

"Timocharis observed these things in Alexandria and reports
that in the 47th year of the first 76-year Kallippic cycle, on the
8th day of the month Anthesterion, according to the Egyptians
the 29th day of the month Athyr, at the end of the 3rd hour,
the southern half of the moon appeared to have advanced toward
the east to the third or half part of the Pleiades, exactly. The
time is during the 465th year of the era of Nabonassar, according to the Egyptians from the 29th to the 30th of the month
Athyr, three seasonal hours before midnight – – – – ."

I Kall. 47, Anth. 8, Athyr 29, end of the third night hour
= Nab. 465, Athyr 29/30 = 283 B.C., Jan. 29, 8:40 P.M.

4. *Almagest*, VII (ed. Heiberg), p. 29: καὶ ἐν τῷ μη´ δὲ ἔτει
τῆς αὐτῆς περιόδου φησὶν ὁμοίως, ὅτι τοῦ μὲν Πυανεψιῶνος τῇ ϛ´
φθίνοντος, τοῦ δὲ Θὼθ τῇ ζ´, τῆς ι´ ὥρας ὅσον ἡμιωρίου προελθόντος
ἐκ τοῦ ὁρίζοντος ἀνατεταλκυίας τῆς σελήνης ὁ Στάχυς ἐφαίνετο ἁ-
πτόμενος αὐτοῦ τοῦ βορείου ἀκριβῶς. καί ἐστιν ὁ χρόνος κατὰ τὸ υξϛ´
ἔτος ἀπὸ Ναβονασσάρου κατ᾿ Αἰγυπτίους Θὼθ ζ´ εἰς τὴν η´, ὡς μὲν
αὐτός φησιν, μετὰ γ̄ Ľ ὥρας καιρικὰς τοῦ μεσονυκτίου – – – .

"And in the 48th year of the same cycle he says that on the
6th day from the end of the month Pyanepsion, on the seventh
day of the month Thoth, at about half-past the ninth hour, the

moon having risen above the horizon, Spica appeared just to touch its northern edge. The time is during the 466th year of the era of Nabonassar, according to the Egyptians from the 7th to the 8th of the month Thoth, as he himself says, at three and a half seasonal hours after midnight – – – –."

    I Kall. 48, Pyan. 24/25, Thoth 7, tenth night hour = Nab. 466, Thoth 7/8 = 283 B.C., Nov. 9, 2:30 A.M.

These four observations, as noted, fix dates by the month and day in the Kallippic cycle, and by the Egyptian month and day, and then add as well the year of the era of Nabonassar and repeat the Egyptian month and day. The Egyptian calendar makes possible the translation into Julian dates. But since the observations were all made at night, and since the Egyptian day began at sunrise, the Greek day at sunset, and the Julian day at midnight, adjustments must be made for the equations in terms of business days, as follows: [6]

|   |   |   |
|---|---|---|
| 1. I Kall. 36, Pos. 25 | = Phaophi 17 | = Dec. 21, 295 B.C. |
| 2. I Kall. 36, Elaph. 15 | = Tybi 6 | = March 10, 294 B.C. |
| 3. I Kall. 47, Anth. 8 [7] | = Athyr 30 | = Jan. 30, 283 B.C. |
| 4. I Kall. 48, Pyan. 24/25 | = Thoth 8 | = Nov. 9, 283 B.C. |

In studying the calendar recently, scholars, though they may disagree with one another in their interpretations, have taken these dates which have names of Attic months to be dates according to the calendar of Athens. Dinsmoor strengthened his belief in the applicability of these astronomical dates to the actual Athenian calendar by citing three additional observations used by Ptolemy in which the names of the Attic months were

---

[6] See W. B. Dinsmoor, *Archons*, p. 365.

[7] These dates were all quoted by Ptolemy from astronomical records at his disposal, and he used them, as he saw fit, in his argument. Somewhere during the course of the compilation an error was made in the writing of Anthesterion in the third equation; the month was Gamelion. Dinsmoor (*Archons*, p. 391 note 2) observes that the error does not affect the astronomical computation, and attributes it to a clerical slip when the Egyptian dates were being converted to Attic (*sic*) equivalents.

accompanied by the names of the Athenian archons.[8] These additional observations were the eclipses of the moon in the years 383 and 382 B.C. Ptolemy reports them under a general heading in which he says that Hipparchos gave them from the Babylonian records, as having been observed in Babylon: [9]

1. *Almagest*, IV (ed. Heiberg), p. 340: γεγονέναι δὲ τὴν πρώτην ἄρχοντος Ἀθήνησι Φανοστράτου μηνὸς Ποσειδεῶνος καὶ ἐκλελοιπέναι τὴν σελήνην βραχὺ μέρος τοῦ κύκλου ἀπὸ θερινῆς ἀνατολῆς τῆς νυκτὸς λοιποῦ ὄντος ἡμιωρίου· καὶ ἔτι, φησίν, ἐκλείπουσα ἔδυ. γίνεται τοίνυν οὗτος ὁ χρόνος κατὰ τὸ τξϛʹ ἔτος ἀπὸ Ναβονασσάρου, κατ᾽ Αἰγυπτίους δέ, ὡς αὐτός φησιν, Θὼθ κϛʹ εἰς τὴν κζʹ μετὰ ε Lʹ ὥρας καιρικὰς τοῦ μεσονυκτίου, ἐπειδήπερ λοιπὸν ἦν τῆς νυκτὸς ἡμιώριον.

" [He says] that the first took place when Phanostratos was archon at Athens, in the month Posideon, and that a small part of the disk of the moon was eclipsed, on the side of the summer rising, while a half hour of the night still remained; and, he says, the moon set while still eclipsed. The time of this was during the 366th year of the era of Nabonassar, according to the Egyptians (as he himself says) from the 26th to the 27th of Thoth, at five and a half seasonal hours after midnight, when one half hour of the night still remained."

2. *Almagest*, IV (ed. Heiberg), p. 341: πάλιν τὴν ἑξῆς ἔκλειψίν φησιν γεγονέναι ἄρχοντος Ἀθήνησιν Φανοστράτου Σκιροφο-

___

[8] W. B. Dinsmoor, *Archons*, p. 350 with note 5, pp. 365, 391.

[9] *Almagest*, IV (ed. Heiberg), p. 340: ταύτας μὲν δὴ τὰς τρεῖς ἐκλείψεις παρατεθεῖσθαί φησιν ἀπὸ τῶν ἐκ Βαβυλῶνος διακομισθεισῶν ὡς ἐκεῖ τετηρημένας — — — — etc. See Dinsmoor, *Archons*, p. 350 with note 3. The eclipses were certainly observed in Babylon: Oppolzer had expressed his skepticism before the publication of the Babylonian tables. The translation of Taliaferro (*Great Books of the Western World*, Vol. 16, p. 140) is incorrect in rendering the Greek "Now he says these three eclipses were given out by those crossing over from Babylon as having been observed there." The Greek means "Now he says that these three eclipses have been adduced from those transmitted from Babylon as having been observed there."

ριῶνος μηνός, κατ' Αἰγυπτίους δὲ Φαμενὼθ κδ′ εἰς τὴν κε′. ἐξέλειπεν
δέ, φησίν, ἀπὸ θερινῆς ἀνατολῆς τῆς πρώτης ὥρας προεληλυθυίας.
γίνεται δὴ καὶ οὗτος ὁ χρόνος κατὰ τὸ τξς′ ἔτος ἀπὸ Ναβονασσάρου
Φαμενὼθ κδ′ εἰς τὴν κε′ πρὸ ε̄ Ľ ὡρῶν μάλιστα καιρικῶν τοῦ
μεσονυκτίου – – – .

"Again he says that the next following eclipse took place
when Phanostratos was archon at Athens, in the month Skirophorion, according to the Egyptians in the month Phamenoth
from the 24th to the 25th day. The moon, he says, was eclipsed
on the side of the summer rising after the first hour of the night
had passed. The time of this too was during the 366th year of
the era of Nabonassar in the month Phamenoth from the 24th
to the 25th day, at approximately five and a half seasonal hours
before midnight – – – – ."

3. *Almagest*, IV (ed. Heiberg), pp. 342-343: τὴν δὲ τρίτην
φησὶν γεγονέναι ἄρχοντος Ἀθήνησιν Εὐάνδρου μηνὸς Ποσειδεῶνος
τοῦ προτέρου κατὰ Αἰγυπτίους Θὼθ ις′ εἰς τὴν ιζ′· ἐξέλειπεν δέ,
φησίν, ὅλη ἀρξαμένη ἀπὸ θερινῶν ἀνατολῶν δ ὡρῶν παρεληλυθυιῶν.
γίνεται δὴ καὶ οὗτος ὁ χρόνος κατὰ τὸ τξζ′ ἔτος ἀπὸ Ναβονασσάρου
Θὼθ ις′ εἰς τὴν ιζ′ πρὸ β̄ Ľ μάλιστα ὡρῶν τοῦ μεσονυκτίου.

"He says that the third took place when Euandros was archon
at Athens in the month of first Posideon,[10] according to the
Egyptians in the month Thoth from the 16th to the 17th day.
It was a total eclipse, he says, beginning on the side of the
summer rising after four hours of the night had passed. And
the time of this was during the 367th year of the era of Nabonassar in the month Thoth from the 16th to the 17th day, at
approximately two and a half hours before midnight."

[10] The translation of Taliaferro (*Great Books of the Western World*,
Vol. 16, p. 141) is incorrect, as is that of Halma in his edition of 1813 with
French translation (p. 278). The reference is to the regular, or non-inter-
calary, Posideon in an intercalary year, not to the first day of the month.

Again the Egyptian dates, as well as the Athenian archons and modern calculations of lunar eclipses, make possible Julian equivalents, and the equations may be tabulated as follows:

Archon Phanostratos, Pos.  = Nab. 366, Thoth 27  = Dec. 23, 383 B.C.
Archon Phanostratos, Skir.  = Nab. 366, Phamenoth 24  = June 18, 382 B.C.
Archon Euandros, Pos. I  = Nab. 367, Thoth 16  = Dec. 12, 382 B.C.

The Athenian dates are not here given as belonging to any cycle, Metonic or otherwise, but by Athenian month and archon. The calendar could be the Athenian festival calendar, as Dinsmoor has argued, in which the archonship of Euandros was intercalary [11] and, presumably, the archonship of Phanostratos ordinary. But it is far easier to refer the dates to the Metonic astronomical cycle, in view of the fact that the observations were made in Babylon. This I believe to be the correct interpretation; and, in any event, the four observations of Timocharis certainly do not belong to the Athenian festival calendar. They do not name an Athenian archon; there is, in fact, no reason to associate the dates in any way with Athens except the names of the months, and these were not named as Athenian. Indeed, as Ptolemy reports them, they are defined by the ordinal number of their year in the first Kallippic cycle. This was the cycle of 76 years (specifically so named in the third equation), a refinement of Meton's 19-year cycle, in which, as Geminus states in his Εἰσα-γωγή, "they made 110 months hollow and 125 full, so that instead of their being alternately hollow and full, there were sometimes two full months in succession." [12] As Pritchett and Neugebauer have so well observed, the "they" to whom Geminus refers were not the Athenians but the astronomers

---

[11] Mention of Ποσειδεῶνος τοῦ προτέρου implies the following intercalated month.

[12] Geminus, VIII, 52 (ed. Manitius, p. 120) [see now Gemini Elementa Astronomiae, ed. E. J. Dijksterhuis, Textus Minores, XXII (Leiden, 1957), p. 37]. One should note here the implication that the Athenians of the fifth century, in their festival calendar, were quite accustomed to ordering their months by a simple process of alternation. See below, pp. 33-36.

Euktemon, Philippos, Kallippos, and their followers.[13] Ptolemy's dates for the observations made by Timocharis were not, therefore, dates in the Athenian calendar but dates in the cycle of Kallippos. The names of the months in the Kallippic year were naturally the Athenian names: Kallippos based his cycle on that of Meton, and Meton propounded his cycle at Athens. As an ideal cycle, there were of course definite rules for the distribution of intercalary months and days—there must have been—and these rules must have been rigorously observed.

All this is quite different from the festival calendar by which the Athenians lived at Athens. Yet these very dates, given in terms of the Kallippic cycle, have been used as evidence for the nature of the Athenian calendar.[14] Dinsmoor, it is true, put forward and then rejected the idea that "the calculations of Ptolemy were based on an ideal astronomical calendar, which need not necessarily have been identical with the civil calendar in regular use at Athens." He thought this rejected interpretation might be acceptable only if we could prove that there were discrepancies between Ptolemy's dates and those in use at Athens, and he believed that no such discrepancies existed.

It so happens, however, that there is disagreement, and that it can be demonstrated. It concerns the calendar character of the two Attic years 295/4 and 294/3 B.C. and the debate as to which of these two years was intercalary and which was ordinary.

In 1938 I restored the one known Athenian equation of the year 294/3 in such a way as to yield an ordinary year: [15]

---

[13] οἱ περὶ Εὐκτήμονα καὶ Φίλιππον καὶ Κάλιππον ἀστρολόγοι. Cf. Pritchett and Neugebauer, Calendars, p. 13.

[14] E. g.: Pritchett and Neugebauer, Calendars, p. 12; W. B. Dinsmoor, Archons, p. 391.

[15] Hesperia, VII, 1938, pp. 98-99.

I.G., II², 378

a. 294/3 a.                              ΣΤΟΙΧ. 33

[ἄρχων Ὀλυμπιόδωρος· ἐπ'] ἀναγραφέως Θρασ

[. . . . . . . . . . . . .¹⁹. . . . . . . . . Φυ]λασίου· ἐπὶ τῆς

[. . .ᵗ. . . . ίδος ἕκτης πρυταν]είας· Ποσιδει

[ῶνος ἑβδόμει μετ' εἰκάδας, τε]τάρτει καὶ ε

[ἰκοστεῖ τῆς πρυτανείας· ἐκκλ]ησία κυρία·

etc.

Posideon [24] = Prytany [VI] 24 [16] = 171st day

The assumption of an ordinary year was taken over by
Pritchett and Meritt in *Chronology*, pp. 87-88 (cf. also *op. cit.*,
p. xvi), where something of a point was made of the fact that
in these first five lines of the text the lettering is strictly stoichedon.[17] Now the suggestion has been made that by restoring
Ποσιδει[ῶνος τετράδι ἱσταμένου] for the date by month in lines
3-4 and [πέμπτης] for the number of the prytany in line 3 an
equation suitable for an intercalary year might be achieved:[18]

Posideon [4] = Prytany [V] 24 = 152nd day.

This recent interpretation of 294/3 as an intercalary year has
been occasioned in part by the desire to make 295/4 ordinary.
In brief, it is claimed that the astronomical observations made
by Timocharis in Alexandria prove that 295/4 had no second
Posideon and hence must be taken as an ordinary year.[19] It
should be noted that the year may have been intercalary by

---

[16] This equation should be read in Pritchett and Neugebauer, *Calendars*,
p. 71, where there is a misprint in the number of the prytany.

[17] I have given line 3 in the text here with the name of the phyle containing
eleven letters. This corrects an error in *Hesperia*, VII, 1938, p. 99 (see the
notes *ad loc.*, p. 98).

[18] Cf. Pritchett and Neugebauer, *Calendars*, p. 71. This would give a text
too short by one letter for the stoichedon requirement of line 4. Pritchett
and Neugebauer (*loc. cit.*) refer to the inscription as "non-stoichedon."

[19] Pritchett and Neugebauer, *Calendars*, p. 72. See also Dinsmoor, *Archons*,
p. 391.

virtue of the intercalation of some month other than Posideon. But this recourse would still not solve the problem of the Egyptian observations:

Posideon 25 = Phaophi 17 = Dec. 21, 295 B.C.
Elaphebolion 15 = Tybi 6 = March 10, 294 B.C.[20]

And these observations (so it has been supposed) yield also the exact dates of the first days of these two Attic months in the year 295/4: [21]

| First Day of Month | | New Moon |
|---|---|---|
| Posideon 1 | = Nov. 27, 295 B.C. | Nov. 26, 7:45 A.M. |
| Elaphebolion 1 | = Feb. 24, 294 B.C. | Feb. 23, 2:30 A.M. |

But the inscriptions have still to be considered.

The epigraphical evidence for the year 295/4 (archonship of Nikostratos) lies in two texts which have been restored in the *Corpus* as follows:

I.G., II², 646

a. 295/4 a. ΣΤΟΙΧ. 30

[θ          ε]        ο        [ι]
['Επὶ Νικοστράτ]ου ἄρχοντος ἐπὶ τῆ[ς Δη]
[μητριάδος ἐνά]της πρυτανείας· Ἐλ[αφη]
[βολιῶνος ἐνάτ]ει ἱσταμένου, πέμπ[τ]ε[ι]
5    [καὶ δεκάτει τῆ]ς πρυτανείας· – – κτλ. –

I.G., II², 647

a. 295/4 a. ΣΤΟΙΧ. 23

['Επὶ Νι]κοστράτου ἄρχοντος [ἐ]
[πὶ τῆς] Δημητ[ρ]ιάδος ἐνάτη[ς π]
[ρυταν]είας ᾗ[ι] Δωρόθεος 'Αρ[ισ]
[τομάχ]ου Φαληρεὺς ἐγρα[μμάτ]
5    [ευεν· 'Ε]λαφη[βο]λιῶνος ἐ[νάτηι]

[20] See above, p. 22.
[21] See above, p. 19.

[ἰσταμ]ένο[υ, πέμπ]τει κ[αὶ δεκ ᵛ]
[άτει τῆς πρυτα]ν[ε]ί[ας ...⁶...]
κτλ.

I have shown here at the end of line 6 in *I.G.*, II², 647, the
uninscribed space which Kirchner recorded only in his commentary. Both inscriptions are dated on the same day, the ninth
of Elaphebolion, which may have been the 15th day of the ninth
prytany, if the restorations in the *Corpus* are followed. But if
the stoichedon order of *I.G.*, II², 647, is to be preferred to that
of *I.G.*, II², 646, then the date within the prytany should be the
25th, with [πέμπ]τει κ[αὶ εἰκο|στεῖ τῆς πρυτα]ν[ε]ί[ας] restored
in *I.G.*, II², 647 (lines 6-7), and πέμπ[τ]ε[ι | καὶ εἰκοστεῖ τῆ]ς
πρυτανείας restored in *I.G.*, II², 646 (lines 4-5). There is no way
to determine, epigraphically, which document preserved perfect
order.[22]

The equation Elaphebolion 9 = Prytany IX [1]5 would be
suitable for an ordinary year, and the equation Elaphebolion 9
= Prytany IX [2]5 would indicate an intercalary year. Because
of the observations of Timocharis at Alexandria the latest suggestions have called for an ordinary year.[23]

There is, however, difficulty with the lengths of prytanies.
Granted an ordinary year, the date within the month should be
the same (or nearly the same) as the date within the prytany.
But, by hypothesis, the ninth of Elaphebolion was the fifteenth
of the prytany, a discrepancy of six days. To avoid undue irregularity in the prytanies one must assume that extra days
(perhaps as many as six) had been intercalated into the festival
calendar. Granted an intercalary year, the 25th day of the ninth
prytany should be the 281st day of the year (approximately),
but Elaphebolion 9 would normally be only the 275th day

[22] Cf. Pritchett and Neugebauer, *Calendars*, p. 71.
[23] Pritchett and Neugebauer, *op. cit.*, p. 72.

(approximately). Hence, again, one must assume the intercalation of extra days (perhaps as many as six). Pritchett and
Neugebauer comment as follows: " As we have seen above,
p. 12, both Posideon (VI) 1 (= 295 November 27) and Elaphebolion (IX) 1 (= 294 February 24) fell on dates which
correspond to the first visibility of the moon. The intercalation
of days must, therefore, have had the purpose of adjusting the
calendar with the moon; hence, this relationship must have been
disturbed at the beginning of the year." [24]

The explanation, whatever it may be, must be different from
this. If the calendar was correct at the beginning of Posideon
by virtue of the extra intercalation of days, then it was not
correct at the beginning of Hekatombaion, the error being
measured by the number of days intercalated. But the thesis
which all students of the calendar maintain, and which is fundamental to the relationship between the festival and conciliar
years, is that the calendars of these two years both be free from
error and begin together at the beginning of each new year. The
only place at which extra days can be assumed is after Posideon
25, which (by hypothesis) was correct according to the Egyptian observation, and before Elaphebolion 9, which would have
been *at least* three days too early for normal prytany lengths
according to *I.G.*, II², 646 and 647.

And the correction, subtracting the same three or more days,
must have been made before Elaphebolion 15, for on that date
the festival calendar was again correct according to the Egyptian observation. The only solution, in fact, that will unite the
evidence of Timocharis at Alexandria with that of the inscriptions at Athens, is to assume that days were added by the archon
to the festival calendar at Athens before Elaphebolion 9 and
again subtracted immediately thereafter.

In the ordinary year posited by Pritchett and Neugebauer

[24] Pritchett and Neugebauer, *Calendars*, p. 72.

the epigraphical equation is

Elaphebolion 9 = Prytany IX 15.

If the last four months had alternately 29 and 30 days, there were left after Elaphebolion 9 still 109 days in the year (with no correction by way of omissions). Hence the last four prytanies must have averaged $(15 + 109) \div 4 = 124 \div 4 = 31$ days each. To bring the average of the prytanies down to a possible 30 days each [25] at least four days must be excised from the festival calendar after Elaphebolion 9 and before Elaphebolion 15. If months and prytanies were running in length *pari passu*, the excision would remove every intervening day.

The interpretation of the evidence, therefore, which associates Athens with the observations of Timocharis, and the consequences of which must now be faced, necessitates dropping from the festival calendar the entire sacred span of the City Dionysia.[26] This is absurd, and it demonstrates the impossibility of using the Egyptian evidence for the reconstruction of the actual calendar at Athens.

What, then, is the meaning of the dates in Posideon and Elaphebolion of 295/4 as given by Timocharis? Since they cannot be Athenian dates we must hold that they are astronomical dates which belong to an ideal calendar used by the astronomers, a calendar which followed fixed rules of months and intercalations, and in which any given date could be equated exactly with its corresponding date in the rigid Egyptian calendar.[27] The existence of this calendar has been maintained by Fotheringham,[28] and should, I think, not be questioned.[29]

---

[25] Pritchett and Neugebauer tolerate no prytany in an ordinary year with more than 30 days in the period of the twelve phylai.

[26] The dates of the festival ran from Elaphebolion 9 through Elaphebolion 13. See L. Deubner, *Attische Feste*, Table facing p. 268.

[27] See above, pp. 25-26.

[28] J. K. Fotheringham, *Monthly Notices of the Royal Astronomical Society*, LXXXIV, 1924, p. 386.

[29] Pritchett and Neugebauer, *Calendars*, p. 72 note 9.

Pritchett and I made a statement in 1941, in which we quoted with approval Fotheringham's views, and which, in my judgment, gives the correct interpretation of these astronomical observations: [30] "Important as are such observations of the ancient astronomers for the astronomical calendar, we wish to emphasize that these dates cannot be used to show the character of any specific year in the civil calendar at Athens. This civil calendar was subject to all kinds of vagaries that might be introduced at the whim of the Athenian demos and can never have been used as a norm for astronomical calculation throughout the Greek world. Although the astronomers used the names of months of the Athenian calendar, their cycles of intercalation must have been the ideal regular cycles of the astronomer. Fotheringham (*Monthly Notices of the Royal Astronomical Society*, LXXXIV, 1924, p. 386), in speaking of calendar equations made by Hipparchos, shows that 'it is out of the question that Hipparchus or any other astronomer should have had the necessary information to convert Babylonian months in distant years into months of the Attic civil calendar, or that succeeding astronomers could have been sure of the dates if they had been so converted. The Babylonian months must have been converted into Greek astronomical months—that is, into months of the calendar used by astronomers to express dates for the time of the observation.....' And again (*Nautical Almanac for 1931*, p. 737), he emphasizes the distinction between astronomical and civil calendars: 'It was an easy matter to compute the interval from one date to another in a calendar regulated by cycle, which was independent of the discretion of city govern-

---

[30] Pritchett and Meritt, *Chronology*, pp. 85-86. I am privileged to say that Neugebauer now finds (in a communication to me) that the observations of Timocharis in Alexandria have no bearing on the calendar in Athens. A discussion of the Athenian lunar month by Pritchett (*Cl. Phil.*, LIV, 1959, pp. 151-157) came to my attention after these lectures were written. It contains nothing, I think, that needs to cause a modification of my arguments in this and in the following chapter.

ments. The original intention may have been merely to facili-
tate the determination of the age of the Moon and the season of
the year, but the Metonic and Callippic cycles at least came to
be used for dating astronomical observations.' "

We base our study of the calendar character of the year
295/4, therefore, solely on the epigraphical evidence. If the year
was ordinary, we must assume that four days or more were
intercalated before Elaphebolion 9, and we are now free to
make the corresponding compensation by omitting four days or
more after the Dionysiac festival was past—at any rate after
Elaphebolion 13.

If the year 295/4 was intercalary, the equation of *I.G.*, II²,
646 and 647 should be Elaphebolion 9 = Prytany IX [2]5.
Now, without correction, the last four prytanies would seem to
have averaged $(109 + 25) \div 4 = 134 \div 4 = 33\frac{1}{2}$ days each. In
order to bring the average length of a prytany down to 32 days
it must be assumed that six days were added before Elaphebolion
9 and subtracted after Elaphebolion 13. The date Elaphebolion
9, which appears on the stone, was in reality, therefore, a date
κατ' ἄρχοντα; the real date κατὰ θεόν was Elaphebolion 15.[31] This
shows major tampering with the festival calendar, but it comes
at a time of year when such tampering is known otherwise to
have taken place, and is not unlike the anomaly of the extra
days in Elaphebolion in 271/0.[32]

The epigraphical evidence favors slightly, I think, the interpretation of 295/4 as intercalary and 294/3 as ordinary.[33] The
observations of Timocharis in Alexandria do not concern us.

When Pritchett and Neugebauer argue against a strict scheme
of alternating months of 29 and 30 days in determining the

[31] For dates κατ' ἄρχοντα and κατὰ θεόν see Pritchett and Neugebauer,
*Calendars*, pp. 14-23.
[32] See W. B. Dinsmoor, *Hesperia*, XXIII, 1954, p. 299, for the text of a
recently discovered inscription from the Athenian Agora. Cf. also W. K.
Pritchett, *B.C.H.*, LXXXI, 1957, pp. 269, 274-275. See below, pp. 151-152.
[33] As shown by Pritchett and Meritt, *Chronology*, p. xvi.

beginnings and the lengths of months, I am in agreement with them.[34] I hold to a scheme of alternating months, for which there is indirect evidence in Geminus's description of the introduction of the Metonic cycle,[35] but I believe that there was nothing compulsory or fixed in advance about it; that is, it was not a strict scheme. From time to time an observation of the new crescent moon (or of any lunar phase) can have been the occasion for adding an extra day. There may have been some incentive to make a controlling observation at the end of the year, which possibly accounts for an extra day from time to time at the end of Skirophorion. But I should not press this point. My disagreement begins where there is insistence on continuous observation of the crescent (for which I think there is no evidence) and when one thus from the lunar year infers a calendar of months "markedly irregular as far as the sequences of 29 and 30 days are concerned."[36] This irregularity is compounded, for—unlike Babylon and Israel—allowance is made for "carelessness in details, resulting in deviations of one or two days, deviations which are at any rate unavoidable in periods of unfavorable weather conditions," and for short-range prediction, where an early full moon, or a date of the last quarter or of the last visibility, might allow a guess as to the date of the coming new moon, even without observation.[37]

In effect, the scheme which my colleagues have advocated, with its various concessions to carelessness and approximation and error, comes no closer to month by month correspondence with the moon than the rule-of-thumb alternation of full and hollow months without observation, or, better said, with only occasional observation. The latter method does not mean "the abolition of all relations of the months of the civil calendar with

[34] Pritchett and Neugebauer, *Calendars*, p. 12.
[35] See above, p. 25 note 12, and below, pp. 35-36.
[36] Pritchett and Neugebauer, *Calendars*, p. 12.
[37] Pritchett and Neugebauer, *Calendars*, p. 13.

the moon," and though it fits pretty accurately the description
of the calendar which Geminus, in his Εἰσαγωγή, attributes to
the astronomers, it was not in any sense ideal, nor was it strict
or rigid, nor was it the system which Geminus was describing.
We do not, indeed, except by indirection, derive our knowledge
of the actual Athenian calendar from Geminus. We must avoid
the application of his rigid ideal scheme to Athens, just as we
must equally avoid the rigidity of a scheme which depends on
continuous observation.

In their practical experience of the calendar the Athenians had,
in fact, hit upon a rule of alternating full and hollow months,
though it remained for the astronomers to codify the rule and
perfect it. But the practice of alternating the months was older
than Euktemon or Philippos or Kallippos. It was older than the
Metonic cycle which Geminus describes, for Geminus based his
description of the new cycle on the older system of alternation.[38]
We are not told how the Athenians managed the occasional
necessary extra day, but they may well have added it, and
reversed the alternation, whenever it seemed desirable to them
to do so.

In the course of time realization must have come that the
average length of the synodic month was 29½ days (or a little
more) [39] and so in the process of time the Athenians came to
avoid juxtaposition of hollow months in succession. These hollow months in succession can be thought to exist in the calendar
only on the assumption, which Pritchett and Neugebauer have
made, of continuous and accurate observation of the new moon
at those times when, astronomically, there were two (or three)
such short synodic lunar months in succession. We have seen
in our study of the year 336/5 B.C. that an assumption of three
short months in succession comes at a time when astronomically

[38] See above, p. 25 note 12, p. 34, and below, p. 36.
[39] Actually *ca.* 29.53 days.

there is no justification for it. If short months are to be posited anywhere in the Athenian calendar in succession and then explained on the theory of observation of the crescent, the least that one can ask is that the short months come where they are astronomically desirable.

If I, in my study of the calendar, use the principle of alternation of full and hollow months in estimating days in the festival year, it is not because I believe that it depends in Athens on any system of astronomical calculation, nor indeed because I believe that it is guaranteed by observation, but because Geminus tells us that it was the traditional Athenian way and because over relatively long periods of time it gives satisfactory results which could be, and probably were, checked by observation whenever it was felt desirable to do so. In this I agree with Pritchett and Neugebauer, holding (as they do) to a festival calendar regulated essentially by the moon, though I place greater reliance than they on the alternation of the months and I do not believe that the beginnings of months were determined by continuous *ad hoc* observation.

There is, indeed, specific evidence from the fifth century on this very point. The beginnings of months were, in fact, known in advance. This will be clear for both the fifth and fourth centuries from our study of the count of days in the following chapter, but for the fifth century there is another very precise indication: in 411 B.C. the length of the month Thargelion was known at least as early as Thargelion 14. Aristotle ('Aθ. Πολ., 32, 1) has this statement about the dissolution of the old Council of the Democracy and the assumption of power by the Four Hundred: ἡ μὲν βουλὴ ⟨ἡ⟩ ἐπὶ Καλλίου πρὶν διαβουλεῦσαι κατελύθη μηνὸς Θαργηλιῶνος τετράδι ἐπὶ δέκα, οἱ δὲ τετρακόσιοι εἰσῄεσαν ἐνάτῃ φθίνοντος Θαργηλιῶνος· ἔδει δὲ τὴν εἰληχυῖαν τῷ κυάμῳ βουλὴν εἰσιέναι τετράδι ἐπὶ δέκα Σκιροφοριῶνος. The Council of the year 412/1 was in its tenth and final prytany when it was dissolved. The number of days in this last prytany was known, as I hold, though I do not claim now to know whether it was to have a normal 36 or 37 days or perhaps show some slight divergence from this norm. But Pritchett and Neugebauer claim that the number of days was surely 36,[40] and it follows for them—as well as for me—that the number of days between Thargelion 14 and Skirophorion 14 was also known. There is no opportunity for choice in calendar reconstruction between making Thargelion full or hollow. Whichever it was, the decision had been reached and the number of its days was known before the middle of the month.[41] The end of Thargelion and the beginning of Skirophorion were not determined, therefore, in 411 B.C. by observation of the lunar crescent. They were determined well in advance by some kind of rule of convenience, which is all that I claim for the alternation of months of 29 and 30 days.[42]

---

[40] Pritchett and Neugebauer, *Calendars*, p. 97.
[41] Or at the latest before ἐνάτη φθίνοντος.
[42] See also below, p. 44.

# CHAPTER III

## The Count of Days

Mention of the date Θαργηλιῶνος ἐνάτη φθίνοντος in 411 B.C. brings to attention once again the question of numbering days in the last decade of a month. It will be well to keep in mind that such days were counted with the addition of a qualifying modifier, either φθίνοντος or μετ᾽ εἰκάδας. Generally speaking, the count with φθίνοντος was used in Athens officially down to the end of the fourth century B.C., the last example now known being from the year 306/5. The count with μετ᾽ εἰκάδας was used officially at Athens during the third and later centuries, but some few examples occur in inscriptions also at the end of the fourth century B.C., the earliest example now being from the year 334/3.

The two systems overlap slightly, but fundamentally they are quite distinct in time just as they are in semantic origin. In spite of this, there has been a tendency to confuse them, and to treat them as though they present only one problem. Scholia, for example, from as late as the fifth century after Christ have been cited as evidence that the omitted day in a hollow month was always δευτέρα φθίνοντος, the 29th. When Pritchett and Neugebauer, for example, in their recent study of the calendar, had reached the conclusion that ambiguity in the meaning of dates could be avoided only by allowing all the days from 21 to 28 inclusive always to have the same designations in both full and hollow months, they found confirmation for this belief in Proklos's comment on a line of Hesiod (*Works and Days*, 765 [Rzach]): ἄρχεται οὖν ὁ Ἡσίοδος ἐκ τῆς τριακάδος, καθ᾽ ἣν ἡ ἀληθής ἐστι σύνοδος, ὁτὲ μὲν οὖσαν τριακάδα ἄνευ ἐξαιρέσεως, ὁτὲ

38

δὲ κθ', ὅτε καὶ ὑπεξαιρεῖται ἡ πρὸ αὐτῆς ὑπὸ 'Αθηναίων.[1] Proklos lived in the fifth century, and was bishop of Constantinople from A.D. 434 to 446. Among his numerous writings were scholia on the *Works and Days* of Hesiod, from which this quotation is taken. But it belongs to an age long after the use of φθίνοντος in Athenian dates within the last decade of a month, being anachronistic in fact by an interval of approximately eight hundred years, and if it has any valid applicability it must refer to the count with μετ' εἰκάδας, not to the count with φθίνοντος.[2]

When a month had thirty days and the count was made in its last decade with φθίνοντος, there is no question about the designation of the days. A firm epigraphical determination is afforded by a series of calendar equations in *I.G.*, I[2], 304B, of the year 407/6:[3]

| Days of the Second Prytany | Days of the Month |
|---|---|
| τρίτ[ει] καὶ δεκάτε[ι] | [δεκάτει φθίνοντος Μεταγειτ]νιῶνος |
| 14 | 9 |
| 15 | 8 |
| 16 | 7 |
| ἑβδό[μ]ει καὶ δεκάτει | [ἔκτει φθίνοντος] Μεταγειτνιῶνο[ς] |
| ἑβδόμει καὶ δεκάτει | ἕ[κτει φθίνοντος Μεταγειτνιῶνος] |
| ὀ[γδ]όει καὶ δεκάτει | [πέμπτει φθίνοντος Μεταγ]ειτνιῶνος |
| ἐνάτει καὶ δεκ[άτ]ει | τε[τ]ράδι φθί[ν]οντος Μετα[γε]ιτνιῶνος |
| 20 | 3 |
| 21 | 2 |
| δευτέραι καὶ εἰκοστ[εῖ] | ἕνει καὶ [ν]έαι Μεταγειτνιῶνος |
| τρίτει καὶ εἰκοστεῖ | [ν]ο[υμε]νίαι Βοεδρομιῶνος |
| | etc. |

[1] Pritchett and Neugebauer, *Calendars*, p. 25. The scholia are now edited by A. Pertusi (Milan, 1955).

[2] See Walther Sontheimer in Pauly-Wissowa, *R.E.*, *s.v.* Monat, col. 52. Later so-called scholars have left at times strange fantasies about the calendar. One commentator on Aristophanes claimed that the Athenians began counting with φθίνοντος on the 15th day διὰ τὸ ἐντὸς ταύτης συμπληροῦσθαι τὴν αὔξησιν τῆς σελήνης καὶ τὴν μείωσιν αὖθις ἄρχεσθαι. For the text see W. J. W. Koster, *Scholia in Aristophanis Plutum et Nubes* (Leiden, 1927), p. 44. The dichotomy of the month reminds one of Hesiod, *Works and Days*, 780: μηνὸς δ' ἱσταμένου τρισκαιδεκάτην.

[3] See the text in B. D. Meritt, *Athenian Financial Documents*, pp. 119-120. This inscription was cited also by A. Mommsen, *Chronologie*, p. 117. The restorations are guaranteed not only mathematically but also by the stoichedon writing of the inscription.

In the fourth century, Demosthenes, in his oration *On the False Embassy* (XIX, 59-60), refers to the twentieth day as εἰκάς, and then names the following days, in order, as ὑστέρα δεκάτη, ἐνάτη, ὀγδόη, ἑβδόμη, ἕκτη, πέμπτη, τετράς. That the modifier φθίνοντος belongs with all these dates from the 22nd to the 27th is shown by his calling the 27th also τῇ τετράδι φθίνοντος.

In the *Clouds* of Aristophanes, Strepsiades is shown making an entrance and muttering to himself the count of days to the end of the month when his creditors will press him for his debts (lines 1131-1134):

πέμπτη, τετράς, τρίτη, μετὰ ταύτην δευτέρα,
εἶθ', ἣν ἐγὼ μάλιστα πασῶν ἡμερῶν
δέδοικα καὶ πέφρικα καὶ βδελύττομαι,
εὐθὺς μετὰ ταύτην ἔσθ' ἔνη τε καὶ νέα.

" The fifth, the fourth, the third, and then the second,
And then that day which more than all the rest
I loathe and shrink from and abominate,
Then comes at once that hateful Old-and-New day."
(Rogers's translation)

The count of days with φθίνοντος was always backward, but it is not clear either from Aristophanes or from Demosthenes whether the months with which they were dealing had 29 or 30 days, for Demosthenes does not carry his count through to the end (leaving us in doubt whether δευτέρα φθίνοντος, perhaps, was omitted) and Aristophanes does not begin his count immediately after the twentieth (leaving us in doubt whether δεκάτη φθίνοντος, perhaps, was omitted). The question of which one of these days was omitted in a hollow month has become one of the tantalizing problems of the calendar.

August Mommsen, in his *Chronologie* (p. 121), cites in part a scholion on line 1131 of the *Clouds* which reads as follows:
τὰς μὲν ἀπὸ νουμηνίας ὁμοίως ἡμῖν, λέγοντες πρώτη ἱσταμένου,

δευτέρα ἱσταμένου, μέχρι τῶν δέκα· μεθ' ἣν πρώτην ἐπὶ δέκα, εἶτα
δευτέραν ἐπὶ δέκα, καὶ ἑξῆς ἄχρι εἰκοστῆς· ἡ αὐτὴ δὲ καὶ εἰκάς·
μεθ' ἣν ἐνδεκάτην φθίνοντος, ἢ δεκάτην, ἢ ἐνάτην, ἢ ὀγδόην, ὡς
ἔτυχεν ἔχων ὁ μήν· καὶ ἑξῆς ἀναλύοντι ἄχρι δευτέρας. οὕτω γὰρ
Ἀθηναῖοι ἠρίθμουν, οὐ προστιθέντες, ἀλλ' ἀφαιροῦντες καὶ ἀναλύον-
τες μέχρι τριακάδος, Σόλωνος ἡγησαμένου, πρὸς τὰ τῆς σελήνης
φῶτα οὕτω μειούμενα. μετὰ δὲ δευτέραν ἀκτέον ἔνην, τουτέστι τὴν
τελευταίαν τοῦ μηνὸς ἡμέραν.

" [they count] the days from new moon as we do, saying
'first of the waxing moon, second of the waxing moon,' as far
as ten; after that 'first on ten,' then 'second on ten,' and so on
in order till the twentieth. This same is also called *eikas*. After
which (they say) 'eleventh of the waning moon,' or 'tenth,'
or 'ninth,' or 'eighth,' depending on the month, and so on with
backward count until the second. For this is the way the
Athenians counted, not adding, but subtracting and counting
backwards until the thirtieth. Solon introduced the scheme,
matching it with the light of the moon which thus diminishes.
After the second one adds the Old (and New), that is, the last
day of the month."

Mommsen would have liked to take this as a definitive proof
that δευτέρα φθίνοντος was not the omitted day, for the count
after the twentieth began with ἐνδεκάτη φθίνοντος, or δεκάτη
φθίνοντος, or ἐνάτη φθίνοντος, or ὀγδόη φθίνοντος, depending
(apparently) on whether the month had 31, 30, 29, or 28 days.
But knowing that the Athenian months alternated between 29
and 30 days he was unable to account for any possible appearance of ἐνδεκάτη φθίνοντος on the one side or of ὀγδόη φθίνοντος
immediately after the twentieth day on the other. It cannot be
claimed that these irregular dates were used when some correction was necessary because of an archon's tampering with the
calendar in one of the first two thirds of the month. We now
have many examples of archons' tampering, most of them too

radical to be corrected by either of the alternatives inferred from the scholion, and, as Mommsen says, there are in fact no known examples of either extreme dating. But rather than take these extremes as products of a scholiast's imagination, pure and simple, I would suggest that he knew the principle of backward count and was misled by thinking of the Athenian month as having the same length as the Roman month. The scholion is clearly of late date—it does not appear in either the Venetus or the Ravennas manuscript of Aristophanes—and a date ἐνδεκάτη φθίνοντος might have seemed appropriate for the 21st day if the month (like January) had 31 days and a date ὀγδόη φθίνοντος might have seemed appropriate for the 21st day if the month had only 28 days (February).[4]

Another late scholion on the *Clouds* of Aristophanes (line 1131) should be quoted here:[5] πέμπτη, τετράς: οὐ προσέθηκε τὸ φθίνοντος. οὕτω γάρ, φασίν, Ἀθηναῖοι μετροῦσιν. ἀπὸ πρώτης ἕως δευτέρας προστιθέασι τὸ ἱσταμένου, ἀπὸ τρισκαιδεκάτης ἕως ἐννεακαιδεκάτης τὸ ἐπὶ δέκα. εἶτα ἡ μεγάλη εἰκάς. ἀπὸ εἰκοστῆς πρώτης ἕως εἰκοστῆς ἐνάτης τὸ φθίνοντος, τὴν δὲ τριακοστήν, ἔνην τε καὶ νέαν, ἐπεὶ μετέχει τοῦ τε παλαιοῦ καὶ τοῦ νέου φωτός. τοῦτον δὲ καὶ Δημητριάδα προσηγόρευσαν. τὴν οὖν εἰκοστὴν δευτέραν ἐκάλουν ἐνάτην φθίνοντος· τὴν εἰκοστὴν τρίτην ὀγδόην· τὴν εἰκοστὴν τετάρτην ζ'· τὴν εἰκοστὴν πέμπτην ϛ'· τὴν εἰκοστὴν ἕκτην ε'· τὴν εἰκοστὴν ἑβδόμην δ'· τὴν εἰκοστὴν ὀγδόην γ'· τὴν εἰκοστὴν ἐνάτην β'· εἶτα τὴν τριακοστὴν ἔνην τε καὶ νέαν, ἐν ᾗ οἱ τόκοι ἀπῃτοῦντο, ἐπεὶ ἐν αὐτῇ συντελεῖται ὁ μήν, καὶ ἀρχὴν ἔχει τοῦ ἑτέρου μηνὸς τῆς σελήνης. ὥστε ἡ μὲν εἰκοστὴ ἕκτη παρ' αὐτοῖς πέμπτη ἀκούει, ἡ δὲ εἰκοστὴ ἑβδόμη τετρὰς λέγεται.

This is a straightforward account of the reckoning of days in an Attic full month. One should read ἀπὸ πρώτης ἕως δεκάτης in

---

[4] The same suggestion was made by W. J. W. Koster, *Scholia in Aristophanis Plutum et Nubes* (Leiden, 1927), plate facing p. 44, and by Adolf Schmidt (cf. Pauly-Wissowa, *R.E.*, *s. v.* Monat, col. 51).

[5] I take the text from Dübner, *Scholia Graeca in Aristophanem* (Paris, 1843).

place of ἀπὸ πρώτης ἕως δευτέρας; designations for the 11th and 12th are omitted; and nothing is said about the 21st. But there is nothing in this scholion to give a clue as to the day omitted in a hollow month.

It is astonishing that the scholia from the lesser manuscripts have been given such prominence, for there are scholia also in the Venetus (eleventh century) and in the Ravennas (tenth century) manuscripts of Aristophanes. These are our oldest manuscripts and the scholia which they afford are the pre-Byzantine "scholia vetera" which go back to the age of Alexandrian scholarship.[6] The publication in the Dübner edition of these valuable scholia can only be called lamentable. From the Venetus I take the first scholion on the *Clouds* (line 1131), transcribing the text from the facsimile publication by John Williams White and Thomas W. Allen (London and Boston, 1902), folio 38b, and I place beside it the text of the Ravennas as copied from the facsimile published by J. van Leeuwen (Leiden, 1904), folio 28 recto.

| VENETUS | RAVENNAS |
|---|---|
| ἐξέρχετ(αι) | ἐξέρχετ(αι) ὁ στρεψιάδ(ης) ἀριθμ(ῶν) τ(ὰς) |
| ὁ στρεψιάδης ἀριθμ(ῶν) | ἡμέρ(ας)· οὕτως δὲ ἠρίθμο(υν) ἀθη |
| τ(ὰς) ἡμέρ(ας)· οὕτως δὲ ἠρίθ | ναῖοι· τὰς μὲν ἀπὸ τῆς νουμη |
| μ(ουν) τ(ὰς) ἡμέρ(ας) ἀθηναῖοι· | νίας ὁμοίως ἡμῖν ἀριθμοῦ |
| τ(ὰς) μὲ(ν) ἀπ(ὸ) τ(ῆς) νουμηνί(ας) ὁ | σι· ᾱ β̄ γ̄ δ̄· ἕως ῑ· εἶτα ῑα |
| μοί(ως) ἡμῖν ἀριθμοῦσι | ῑβ· τρίτη ἐπιδέκα· τετάρτη |
| ᾱ β̄ γ̄ δ̄ ἕως ῑ {τρίτης} | ἐπιδέκα. ἕως εἰκάδος. εἶτα |
| εἶτα ῑα ιβ ἕως τρίτη(ς) | ἣν ἡμεῖς εἰκάδα πρώτ(ην) |
| ἐπιδέκα· τετάρτη ἐ | ἐννάτ(ην) φθίνοντο(ς) αὐτοί φ(ασι)· |
| πὶ δέκα ἕως εἰκάδος· | εἶτα ὀγδόην ἕως ἔν(ης) τε |
| εἶτα ἣν ἡμεῖ(ς) κ̄α ἐνά | κ(αὶ) νέας. |
| τ(ην) αὐτοὶ φα(σιν)· εἶτα ὀγδόη(ν) | |
| ἕως ἔνης τ(ε) κ(αὶ) νέας. | |

This is an explicit statement, substantially the same in both

---

[6] On the scholia see Konrad Zacher, "Die Handschriften und Classen der Aristophanesscholien," *Jahrbücher für classische Philologie*, Suppl. XVI, 1888, pp. 501-746; John Williams White, in his introduction to the *Scholia on the Aves of Aristophanes* (Boston and London, 1914).

manuscripts, giving the count of days in a month of 29 days and making it clear that the omitted day was not δευτέρα φθίνοντος, but rather δεκάτη φθίνοντος.[7] It is also the clinching and formal proof of the argument advanced in our preceding chapter that the length of a month was known before its end and was not determined (on the last day) by observation of the new lunar crescent.

Dübner (*op. cit.*) published the concluding lines of the scholion as follows: εἶτα ἦν ἡμεῖς εἰκάδα πρώτην, δεκάτην φθίνοντος αὐτοί φασιν, εἶτα ἐνάτην ἕως ἕνης τε καὶ νέας. But this is not the reading of the manuscripts. Dübner was aware of this, for he noted in his *apparatus criticus* that the Venetus manuscript has ἐνάτην, the Ravennas ἐννάτην, and the Marcianus G likewise ἐννάτην, all of which he prints as δεκάτην. For the 22nd day ὀγδόην is the reading of the manuscripts; Dübner printed ἐνάτην. The errors thus gratuitously introduced into the published text go back to emendations made by Dindorf, who introduced his "corrections" from another scholion— of less worth—which described a full month. Apparently Dindorf wanted to make all the texts say the same thing, and he thought he knew what that ought to be![8]

There is another scholion, introduced by the lemma μετὰ ταύτη(ν) δευτέρα, which comes from the Ravennas and Venetus manuscripts: τὴν δευτέραν κ(αὶ) εἰκάδα ἐν[άτην] φθίνοντο(ς) ἀθηναῖοι [ἐκάλουν] κ(αὶ) τ(ὴν) τρίτ(ην) ὀγδόην κ(αὶ) [τ(ὴν) τετάρτην] ἑβδόμην κ(αὶ) τ(ὴν) πέμπτ(ην) ἕκτ(ην) · κ(αὶ) τ(ὴν) ἕκτ(ην)

---

[7] August Mommsen cites also as evidence that δευτέρα φθίνοντος was not omitted the statement of Pollux (VIII, 117): καθ᾽ ἕκαστον δὲ μῆνα τριῶν ἡμερῶν ἐδίκαζον (οἱ Ἀρεοπαγῖται) ἐφεξῆς, τετάρτῃ φθίνοντος, τρίτῃ, δευτέρᾳ. If the statement is taken at its face value, there was a day designated δευτέρα φθίνοντος in every month, hollow as well as full. See Pauly-Wissowa, *R.E.*, *s.v.* Monat, col. 51.

[8] Dindorf's arbitrary "corrections" were protested by W. J. W. Koster, *Scholia in Aristophanis Plutum et Nubes* (Leiden, 1927), table facing p. 44, and he rightly concluded that the equation of the 21st day with ἐνάτη φθίνοντος defined a hollow month.

πέμπτ(ην) κ(αὶ) τ(ὴν) ἑβδόμην τετάρτ(ην)· κ(αὶ) τ(ὴν) ὀγδόην
[τρίτην κ(αὶ) τ(ὴν)] ἐνάτην κ(αὶ) εἰκάδα β φθίνοντος, εἶτα τὴν
τριακάδα, ἔνην τε κ(αὶ) ν[έαν].[9]

This description belongs to a full month, but carries the
count only from the 22nd day to the end, presumably because
the 21st day was no longer thought of as δεκάτη φθίνοντος, but
rather as δεκάτη ὑστέρα. This terminology (δεκάτη ὑστέρα)
was attested in Demosthenes for the year 344/3, and appears
epigraphically as early as 302/1 (I.G., II², 505). This brings its
official use in chancery style to a date when the count in the
last decade with φθίνοντος had given way to the count with
μετ' εἰκάδας, a distinction which the scholiast may well not have
made.

The days of the month may now be tabulated, down to the
end of the fourth century, as follows:[10]

DAYS OF THE MONTH IN CHANCERY STYLE DOWN TO THE
LAST DECADE OF THE FOURTH CENTURY B.C.

| Every Month | Every Month | Full Month | Hollow Month |
|---|---|---|---|
| 1. νουμηνία* | 11. ἐνδεκάτη* | 21. δεκάτη φθίνοντος* | 21. ἐνάτη φθίνοντος |
| 2. δευτέρα ἱσταμένου* | 12. δωδεκάτη* | 22. ἐνάτη φθίνοντος | 22. ὀγδόη φθίνοντος |
| 3. τρίτη ἱσταμένου | 13. τρίτη ἐπὶ δέκα | 23. ὀγδόη φθίνοντος | 23. ἑβδόμη φθίνοντος |
| 4. τετρὰς ἱσταμένου* | 14. τετρὰς ἐπὶ δέκα* | 24. ἑβδόμη φθίνοντος | 24. ἕκτη φθίνοντος |
| 5. πέμπτη ἱσταμένου | 15. πέμπτη ἐπὶ δέκα | 25. ἕκτη φθίνοντος | 25. πέμπτη φθίνοντος |
| 6. ἕκτη ἱσταμένου | 16. ἕκτη ἐπὶ δέκα | 26. πέμπτη φθίνοντος | 26. τετρὰς φθίνοντος |
| 7. ἑβδόμη ἱσταμένου | 17. ἑβδόμη ἐπὶ δέκα | 27. τετρὰς φθίνοντος | 27. τρίτη φθίνοντος |
| 8. ὀγδόη ἱσταμένου | 18. ὀγδόη ἐπὶ δέκα | 28. τρίτη φθίνοντος | 28. δευτέρα φθίνοντος |
| 9. ἐνάτη ἱσταμένου | 19. ἐνάτη ἐπὶ δέκα | 29. δευτέρα φθίνοντος | 29. ἕνη καὶ νέα |
| 10. δεκάτη ἱσταμένου | 20. εἰκοστή* | 30. ἕνη καὶ νέα | |

---

[9] This text is taken from the facsimile publication of the Ravennas manuscript by J. van Leeuwen (Leiden, 1904). See also the apparatus criticus in Rutherford's Scholia Aristophanica, I (1896), pp. 249-250. The text in Dübner's edition is misleading.

[10] One should not expect a sharp line to be drawn between this method of dating and that in which the phrase μετ' εἰκάδας appears. See above, p. 38, and below, pp. 57-58. There is brief comment in the notes which follow the table on those dates that are marked with an asterisk.

46 The Count of Days

1. For νουμηνία see Pritchett and Neugebauer, *Calendars*, p. 31. The reference to *I.G.*, I², 304, should be to *I.G.*, I², 304B, lines 53-54 (= B. D. Meritt, *Athenian Financial Documents*, p. 119). Reference to the appearance of νουμηνία in the Assessment Decree of 425 B.C. may now be made to Meritt, Wade-Gery, McGregor, *The Athenian Tribute Lists*, II (1949), p. 41 (A9 line 19).

2. I know of no instance in which the modifier ἱσταμένου is omitted in the preamble of an Attic decree, though the simple δευτέραι (without ἱσταμένου) is given as the second day (of Boedromion) in *I.G.*, I², 304B, line 55 (= B. D. Meritt, *Athenian Financial Documents*, p. 119).

3. For τετρὰς ἱσταμένου, never τετάρτη ἱσταμένου, see Pritchett and Neugebauer, *Calendars*, p. 31.

4. ἐνδεκάτη and δωδεκάτη are the designations for the 11th and 12th days of the month. The Venetus scholion on Aristophanes, *Clouds* (line 1131), makes this clear in defining the days after the tenth as εἶτα ῑα ῑβ ἕως τρίτη(s) ἐπιδέκα. The Ravennas scholion is equally precise: εἶτα ῑα ῑβ· τρίτη ἐπιδέκα. It was wrong of Rutherford to expand ῑα and ῑβ in his published text to πρώτη ἐπὶ δέκα and δευτέρα ἐπὶ δέκα. See August Mommsen, *Chronologie*, pp. 94-96, and W. J. W. Koster, *Scholia in Aristophanis Plutum et Nubes* (Leiden, 1927), table facing p. 44, for the evidence of other scholia and of the grammarians. The phrases πρώτη ἐπὶ δέκα, μία ἐπὶ δέκα, and δευτέρα ἐπὶ δέκα are foreign to Athenian chancery style.

5. As in the case of τετρὰς ἱσταμένου for the 4th day, the correct term for the 14th day uses τετράς. The form τετάρτη ἐπὶ δέκα was not used epigraphically.

6. For εἰκοστή as the 20th day see Pritchett and Neugebauer, *Calendars*, pp. 31-32. Although the phrase δεκάτη προτέρα, or its equivalent δεκάτη προτεραία, is known epigraphically as early as *ca.* 327/6 B.C. (*I.G.*, II², 1673, line 77) I have not shown it here in the table. Along with δεκάτη ὑστέρα for the 21st day, it belongs in very large measure (at least) to the era when days at the end of the month were counted with μετ᾽ εἰκάδας. It is not, however, to be excluded as a possibility. When δεκάτη προτέρα meant 20th day and δεκάτη ὑστέρα meant 21st day, δεκάτη ὑστέρα was not omitted in the hollow month. Evidence for this is now at hand in a decree, recently found, from the archonship of Pytharatos (271/0 B.C.). See below, pp. 194-195. The use of εἰκάς for the 20th day is attested in the literary tradition, and in an unpublished religious decree (Inv. No. I 6108) of 111/0 B.C. from the Athenian Agora.

7. For the days from 21 to 30 see above, pp. 38-45. There was no τετάρτη φθίνοντος in the chancery style, only τετρὰς φθίνοντος. When δεκάτη ὑστέρα replaced δεκάτη φθίνοντος as the designation of the 21st day, this was held even in a hollow month (see below, p. 195). If the rest of the month was still using the count with φθίνοντος, presumably ἐνάτη φθίνοντος was the date omitted when the month was hollow.

There is no ambiguity about the count of days as shown above in the last decade of a month. On the contrary, the Athenian knew precisely, from the very name of the day, how many days were left before the end of the month, this being the important consideration for him, for on that day he had to settle his debts and satisfy his creditors.[11] This was what worried Strepsiades. How much more worried might he not have been had he been compelled to face the omission of his last possible day of grace, the δευτέρα φθίνοντος.

We are guilty of imposing modern concepts of thought on the ancient Athenians if we find that these ordinal differences (in English) in the progressive count through the month are ambiguous. The situation may at first seem puzzling to us. But the Athenian preferred to know how far he was from the end of the month, not from its beginning. When objection is made that then "the character of a month as full or hollow must always have been decided in advance, in order to know how to denote the days from the 22nd onwards,"[12] we realize that just this is exactly what had been decided; when the objection is further made that this requirement "practically excludes a real lunar calendar,"[13] we realize that a rule of convenience in alternating full and hollow months, with occasional corrections,

---

[11] This necessity should be remembered in connection with the calendar, as well as the regulation of more public transactions, sales, leases, etc., which were ordered by the conciliar calendar. See Pritchett, B.C.H., LXXXI, 1957, p. 273.

[12] Pritchett and Neugebauer, Calendars, p. 24.

[13] Pritchett and Neugebauer, Calendars, p. 24.

48   The Count of Days

gave a workable calendar as near to the real lunar calendar as the Athenians thought worth the trouble to achieve.

We have still to turn attention to the epigraphical evidence. Confirmatory proofs lie in the inscriptions of the year 333/2 B.C. First, the preamble of *I.G.*, II², 338, shows that the first prytany had 39 days, that Hekatombaion was full, and that the last day of the prytany fell on Metageitnion 9:

*I.G.*, II², 338

a. 333/2 a.                               ΣΤΟΙΧ. 35

θ            ε            ο            ι
'Επὶ Νικοκράτους ἄρχοντος ἐπὶ τῆς Αἰγηίδος
πρώτης πρυτανείας ἧι Ἀρχέλας Χαιρίου Παλ
ληνεὺς ἐγραμμάτευεν· Μεταγειτνιῶνος ἐνά
5   τηι ἱσταμένου, ἐνάτηι καὶ τριακοστῆι τῆς ᵛ
πρυτανείας· τῶν προέδρων – – κτλ. – – – –

Another inscription comes from the second prytany but from the same month:

*I.G.*, II², 339 [14]

a. 333/2 a.                               ΣΤΟΙΧ. 25

['Επὶ Νι]κοκράτου[ς ἄρχοντος ἐπὶ]
[τῆς Πα]νδιονίδ[ος δευτέρας πρυ]
[τανεί]ας ἧι ἐγρ[αμμάτευεν Ἀρχέ]
5   [λας Χα]ιρίου Πα[λληνεύς· Μεταγε]
[ιτνιῶ]νος ἔκτ[ηι φθίνοντος, πέμ]
[πτηι κα]ὶ δε[κάτηι τῆς πρυτανεί]
[ας – – – – – – – – – – – – – – –]

These two inscriptions give the following calendar equations:

---

[14] See the note on this text in Pritchett and Neugebauer, *Calendars*, p. 48 note 27.

Fig. 1. *I.G.*, II², 339: the first seven lines

Metageitnion 9    = Prytany I 39        = 39th day

Metageitnion [2]4 = Prytany [II] 1[5] = 54th day.

The second text was treated at length some years ago by me and now has been more recently studied by Pritchett and Neugebauer.[15] There is still more to be said, for in their wish to avoid the calendar equation [Μεταγειτνιῶ]νος ἕκτ[ηι φθίνοντος πέμπτηι κα]ὶ δε[κάτηι τῆς πρυτανείας], which proves that δευτέρα φθίνοντος was not omitted in the count of days in a hollow month, our latest investigators have suggested that the reading and restoration may have been [Μεταγειτνιῶ]νος ἕκτ[ηι ἐπὶ δέκα· ἐκκλησία κυρ]ία ἐ[ν – – – –]. This cannot be. The letters ΙΔΕ are all clear in line 7 and the letter Δ cannot legitimately be read as alpha. Pritchett and Neugebauer describe it as follows: "The cross-bar is clearly discernible on the squeeze in such a position that it is overlapped by the oblique hastas. This cross-bar is not placed as high as those in the certain alphas of this inscription, nor is it inscribed on the line so as to make a perfect isosceles triangle, as is the case with the preserved deltas in line 3. Epigraphically, the letter may be read more probably as a delta, but possibly as an alpha."

A photograph of this fragment appears here in Figure 1. One should observe that the alphas spread more widely at the bottom than do the deltas (including this in line 7) and that the free-standing ends of the oblique strokes in alpha are splayed. There is no thickening of the end of any stroke in delta, for there was no free-standing end. It is an understatement to say that the cross-bar in line 7 is not placed as high as those in the certain alphas. The cross-bar in the certain alphas is remarkably high, and short; the cross-bar in line 7 is practically at the bottom, and long. The splaying of the free strokes is quite noticeable not

---

[15] B. D. Meritt, *Hesperia*, IV, 1935, pp. 533-534; Pritchett and Neugebauer, *Calendars*, pp. 46-48.

only in the alphas of this text but also in other letters, all three
of the strokes of the upsilon of line 5, for example, being almost
cuneiform in appearance. It does not help to cite here examples
of malformed alphas in other inscriptions,[16] for this is a well-
formed delta, which no one would interpret otherwise if he
were free of compulsion, and in which the strokes are correct
for delta (both collectively and individually) but not for alpha.

So we are back with the problem of the calendar equation
and the evidence it has to offer. If one writes the date by
prytany as [ἕκτηι κα]ὶ δε[κάτηι] [17] an irregularity must be assumed in an otherwise blamelessly stoichedon text: there must
have been two uninscribed spaces at the end of line 6 after the
date by month ἕκτ[ηι φθίνοντος], or one uninscribed space if
the date is restored as ἕκτ[ηι μετ' εἰκάδας], which was not the
normal method of designating days at this time. Instead of
holding to rigid lengths of prytanies or to a discredited system
of counting days in a month, I prefer here to follow the evidence
of the inscription, and I see no reason to change the opinion that
I expressed about it in 1935: [18] "The restoration in *I.G.*, II²,
339 is: [Μεταγειτνιῶ]νος ἕκτ[ηι φθίνοντος πέμπτηι κα]ὶ δε[κάτηι
τῆς πρυτανείας - - -]. The restoration itself is inevitable, and
may be accepted as correct."

There are two alternatives. If ἕκτη φθίνοντος here means the
25th day of the month, then this was the 55th day and the first
prytany had 40 days. If ἕκτη φθίνοντος can mean the 24th day
of the month (the month being hollow), then the month here
was hollow and the first prytany had a normal 39 days. Surely
the latter is the correct solution, especially since the preceding
month was full. The phrase ἕκτη φθίνοντος means exactly what
it says and what the scholia on Aristophanes say that it says:

[16] See Pritchett and Neugebauer, *Calendars*, p. 47 note 26.
[17] See Pritchett and Neugebauer, *Calendars*, pp. 46-47.
[18] *Hesperia*, IV, 1935, p. 532.

the sixth day from the end of the month, whether that month be full or hollow. Since both the lengths of prytanies and the alternation of full and hollow months in this case require that Metageitnion be hollow, I so restore it and read the date as the 24th of the month. The epigraphical and literary evidences are in agreement: in a hollow month it was not the day before the last (δευτέρα φθίνοντος) that was omitted, but the day with which the count with φθίνοντος would have begun, had the month been full.[19]

This was the first problem of the count of days at the end of a month. The second problem is more complex, and concerns the count when μετ᾽ εἰκάδας, rather than φθίνοντος, was used by way of designation. Part of the complexity is occasioned by the semantic value of the count μετ᾽ εἰκάδας as a forward count. The second day μετ᾽ εἰκάδας was the 22nd, the third day μετ᾽ εἰκάδας was the 23rd, and so forth. This direction of count was contaminated, however, by the habit of using φθίνοντος, so that the phrase μετ᾽ εἰκάδας lost its semantic value and came to be used as the equivalent in meaning of φθίνοντος. Thus in a full month δευτέρα μετ᾽ εἰκάδας came to mean the 29th day, τρίτη μετ᾽ εἰκάδας the 28th day, and so on. But the semantic value was never wholly obscured, and we do in fact have to reckon with forward count even in the epigraphical texts.

This is a disputed point, but after all the circumstantial evidence that can be otherwise explained has been explained away there remains a hard core which can be satisfied with nothing but forward count. Even those who deny the use of forward count most vigorously have been willing to suggest it on rare occasions.[20] But if forward count is to be admitted at all, then

---

[19] It will be evident that much of the *a priori* speculation (*B.C.H.*, LXXXI, 1957, p. 283) about a "much simpler and more plausible" system does not agree with the evidence and is, in spirit, not Athenian.

[20] Cf. Pritchett and Neugebauer, *Calendars*, p. 53.

we must be prepared to test every appearance of the phrase for the possible, or probable, direction of its count.

An inscription of the year 302/1 is restored to read as follows: [21]

### I.G., II², 504

a. 302/1 a.                 ΣΤΟΙΧ. 26

[θ      ε]      ο        ί

['Επὶ Νικοκλέους ἄρχ]οντος ἐπὶ τῆ

[ς Αἰαντίδος δωδεκάτ]ης πρυτανε

[ίας ἧι Νίκων Θεοδώρο]υ Πλωθεὺς ἐ

5 [γραμμάτευεν· Σκιροφ]οριῶνος δε

[υτέραι ⟨μετ'⟩ εἰκάδας, δευτέ]ραι καὶ εἰ

[κοστεῖ τῆς πρυτανείας·] ἐκκλησί

etc.

Since the year is known to have been ordinary, Skirophorion 22 is equated with Prytany XII 22. The equation is guaranteed by another text of the same year and month and prytany (*I.G.*, II², 505) in which Skirophorion 21 is equated with Prytany XII 21. The error of omission of μετ' in the phrase μετ' εἰκάδας is intelligible in *I.G.*, II², 504; at least, it is a simple scribal error. If we wish to avoid forward count we may assume that the date was simply wrongly given;[22] but this involves more than a scribal error, and would be thought necessary, I believe, only if one wanted at any cost to have no forward count. If the value of the numismatic evidence is sound (to be recorded below, pp. 180-191), then there is also a clear case of forward count in the year 137/6 where the date by month [Θαργη]λιῶνος τρίτει μ[ε]τ' εἰκάδας must be counted as the 23rd day of the month.

---

[21] The first line [θε]οί is not recorded in the *Corpus*.

[22] Pritchett and Neugebauer, *Calendars*, p. 27, have noted that δε[κάτει προτέραι, δευτέ]ραι καὶ εἰ[κοστεῖ] would not violate stoichedon order and would assume no omission. It would, however, have the remarkable consequence of bringing the 20th day of Skirophorion to the day following the day which it precedes.

There are also equations which with forward count show no irregularities, and which with backward count require the assumption of tampering with the festival calendar. These occur in 249/8 (*I.G.*, II², 680),[23] in 222/1 (below, p. 173), in 188/7 (below, pp. 154-158), in 178/7 (below, pp. 158-159), and in 160/59 (below, pp. 162-163). There may be other years where a preference could be expressed, but which we interpret with backward count because this was the usual choice.

There is little good independent evidence for the omitted day. But if prytanies in the year 307/6 are to be kept to even lengths (as seems desirable) then the omitted day in the hollow month was the same with μετ' εἰκάδας as it was with φθίνοντος when the count was backward.[24] Since δεκάτη ὑστέρα now regularly meant the 21st day, the omitted designation in a hollow month was ἐνάτη μετ' εἰκάδας. The same holds when the count was forward, for the count would progress normally toward ἕνη καὶ νέα (the last day of the month) and if that day, ἕνη καὶ νέα, was in fact the 29th day, the day before it was ὀγδόη μετ' εἰκάδας and no day was named ἐνάτη μετ' εἰκάδας until the following full month.

Thus it appears that when the count was forward one could say that the day before the last day was the omitted day: this may have been what Proklos had in mind when he wrote the scholion on Hesiod quoted above (pp. 38-39).[25] If he was writing of the older count with φθίνοντος, we know that he was in error. If he was writing of the backward count with μετ' εἰκάδας he was probably in error. If he was speaking of the calendar of his own time (he uses the present tense in describing Athenian practice) we have no evidence with which to control his statement.

[23] G. Klaffenbach, in *Gnomon*, XXI, 1949, p. 135, has protested that here not to adopt forward count "heisst doch wirklich, die Augen gewaltsam verschliessen."

[24] See below, pp. 176-177.

[25] This was the view taken by August Mommsen, *Chronologie*, pp. 122-123. Cf. Walther Sontheimer, in Pauly-Wissowa, *R.E.*, *s. v.* Monat, col. 52

Perhaps the designations of days had all so lost their semantic value by that time that they were nothing more than stereotyped symbols, like those in the count of a calendar from Rhodes which has been adduced for the direction of the count in Athens and for the omitted day. In both full and hollow months at Rhodes in Roman times κ̄ᾱ was always the 21st, but κθ′ was the 22nd, κη′ the 23rd, and so on until κγ′ which was always the 28th, and then the 29th was either πρ(οτριακάς) or τρ(ιακάς) depending on whether the month was full or hollow.[26] The Rhodian text gives the last day as τρ(ιακάς), which has of course a false semantic value in a 29-day month. But all these designations are purely formal and have lost their original meanings. Their applicability to what happened in Athens in the centuries before Christ is remote both in time and place, and to be accepted would need some confirmation from an Athenian source. Moreover, I suspect that the Rhodian letters should not be expanded into the calendar terminology with μετ᾽ εἰκάδας which was in Attic inscriptions, but that κθ′, for example, should be read as εἰκάδα ἐνάτην (or εἰκὰς ἐνάτη) like the count of days in a late scholion on Demosthenes (XIX, 57): εἶτα λοιπὸν ἀπὸ τῆς α′ καὶ εἰκάδος ἐξ ὑποστροφῆς (οἱ Ἀθηναῖοι ἀριθμοῦσι), λέγοντες τὴν α′ καὶ εἰκάδα ὑστέραν δεκάτην, καὶ τὴν β′ εἰκάδα ἐνάτην, καὶ τὴν γ′ εἰκάδα ὀγδόην, καὶ ὁμοίως οὕτως ἕως τριακάδος.[27] We are here told that the Athenians called the 21st day ὑστέρα δεκάτη (which we know epigraphically to have been δεκάτη ὑστέρα), and then that the 22nd day was called εἰκὰς ἐνάτη, the 23rd day εἰκὰς ὀγδόη, etc., down to the last day which was called τριακάς. These designations are not epigraphically attested at Athens, but they would suit admirably the calendar at Rhodes.[28]

---

[26] *I.G.*, XII, 1, 4. Cf. Pritchett and Neugebauer, *Calendars*, p. 25.

[27] Backward count is here described without reference to the hollow month.

[28] One should note that Strepsiades, in the *Clouds* of Aristophanes, calls the last decade of the month εἰκάδας: ἐγὼ δ᾽ ἀπόλλυμαι, ὁρῶν ἄγουσαν τὴν σελήνην εἰκάδας (lines 16-17). A scholiast reports that he means not merely the 20th day but all the days from the 20th down to the 29th.

At the risk of raising a new problem rather than contributing to the solution of an old one I should like to call attention to a difficult text of the year 318/7 B.C., which I have restored below (p. 126) as follows:

*I.G.*, II², 448, lines 35-38

*a.* 318/7 *a.* ΣΤΟΙΧ. 41

Ἐπὶ Ἀρχίππου ἄρχοντος ἐπὶ τῆς Ἀκ[αμαντίδος τετάρ]
της πρυτανείας ἦι Θέρσιππος Ἱππο[...⁶... Κολλυτε]
ὺς ἐγραμμάτευε· Μαιμακτηριῶνος ἕ[κτει ⟨μετ'⟩ εἰκάδας, πέμ]
πτει καὶ τριακοστεῖ τῆς πρυτανεία[ς· τῶν προέδρων]
κτλ.

Maimakterion [25] = Prytany IV 35 = 143rd day

Since the other equations of the year give satisfactory correspondences between the festival and conciliar years, with no irregularities in either, this equation also ought to fall into place in an ordinary year. No supplement suggested hitherto allows it to do so without the assumption of some gross irregularity.

The restoration adopted by Kirchner, Μαιμακτηριῶνος [ἕνει καὶ νέαι, πέν]πτει καὶ τριακοστεῖ τῆς πρυτανεία[ς], is the only one with a known, or accustomed, formula of date within the month that permits retention of the stoichedon order. Kirchner assumed an intercalary year, with Prytanies I and II of 38 days and Prytany III of 37 days. With an ordinary year, the 35th day of the fourth prytany should be the 143rd day of the year (or nearly so), and yet the last day of Maimakterion comes to the 147th or (better) the 148th day of the year. Had the date by month been earlier than the date by prytany, one might have found refuge in the assumption that five extra days had been intercalated somewhere before the end of Maimakterion, giving a date κατ' ἄρχοντα at variance with the true date κατὰ θεόν. But with the date by month later than the date by prytany this approach to a solution is barred: there is no known example of a date κατ᾽ ἄρχοντα being later than its date κατὰ θεόν.

One solution has been to restore Μαιμακτηριῶνος [ἕκτει ἐπὶ δέκα, πέν]πτει καὶ τριακοστεῖ τῆς πρυτανεία[ς].[29] This involves a violation of the stoichedon order, for the calendaric phrase ἕκτει ἐπὶ δέκα is longer by one letter than ἕνει καὶ νέαι. But this is not prohibitive, and in this instance Kirchner's remark has been cited: "Ordo στοιχηδόν saepius turbatus est."[30] As every epigrapher knows, the stoichedon order is never a formal bar to a good restoration, even if a violation has to be assumed in an otherwise perfect text; but here the 16th day of Maimakterion does not in fact equal the 35th day of the fourth prytany. Our latest investigators have assumed, therefore, that the word τριακοστεῖ was written by mistake for εἰκοστεῖ,[31] arriving at the equation

Maimakterion [16] = Prytany IV ⟨2⟩5 = 133rd day.

If the date by prytany as given in the inscription is held to be accurate, the problem resolves itself into finding a restoration of the date by month which will permit a normal calendar. The 35th day of the fourth prytany ought to have fallen on Maimakterion 25 (the 143rd day). Yet the usual formulae for Maimak-

---

[29] Pritchett and Neugebauer, *Calendars*, p. 64. But see below, p. 206 note 11. If one may assume that five days had been omitted from the festival calendar shortly before Maimakterion ἕνη καὶ νέα, that restoration (epigraphically irreproachable) could still be made and equated with Prytany IV, 35, as the 143rd day. Presumably then five ἐμβόλιμοι ἡμέραι would have been written into the calendar by way of compensation soon after.

[30] In my judgment, Kirchner's remark is too strong. The στοιχηδόν order is rarely violated, and in the immediate context of this passage it is strictly preserved.

[31] Pritchett and Neugebauer, *Calendars*, p. 64: "All editors assume that δεκάτηι was inscribed by error for εἰκοστῆι in *I.G.*, II², 351 + 624, and we think it not unlikely that a numerical notation in the scribe's copy of *I.G.*, II², 448B may have resulted in the stonecutter's writing of τριακοστεῖ for εἰκοστεῖ." For my belief that δεκάτηι was correctly written in *I.G.*, II², 351, see below, pp. 92-93. But for an error resulting from a misreading of numerical notation, see also below, p. 161.

terion 25 (ἕκτει φθίνοντος, πέμπτει μετ᾽ εἰκάδας, ἕκτει μετ᾽ εἰκάδας, πέμπτει φθίνοντος) are all too long for the available space of eleven letters.[32] The phrase ἕκτει μετ᾽ εἰκάδας is too long by four letters. If the preposition μετ᾽ was omitted, ἕκτει εἰκάδας would be too long by one letter, but might have as good a claim as ἕκτει ἐπὶ δέκα. But the phrase written simply as ἕκτει εἰκάδας could not be justified; it would have to be written, even in restoration, as ἕ[κτει ⟨μετ᾽⟩ εἰκάδας], and this is the way I have given it in the text above (p. 55).

There is, however, another supplement that one should perhaps mention. The scholiast on Demosthenes (XIX, 57) whose note was quoted above (p. 54) named the days near the end of a month, implying that the Athenians referred to the 25th day as εἰκὰς ἕκτη. May not the date at some time actually have appeared so? The restoration Μαιμακτηριῶνος ε[ἰκάδι ἕκτηι] in I.G., II², 448, line 37, would conform perfectly to stoichedon requirements and would meet admirably the calendar necessity for the 25th day of the month. If the Rhodian text mentioned above (p. 54) could be taken as evidence for Athens, which I find difficult, it would give support to the reading εἰκάδι ἕκτηι. After all, the phrase μετ᾽ εἰκάδας was just coming into official use at Athens in the latter part of the fourth century and the precise form of it may not have become settled by 318/7 B.C. The earliest instance known is from 334/3 (I.G., II², 335), and with the present instance included there are only six examples earlier than 307/6. If we consider this example before us in I.G., II², 448, together with that from I.G., II², 504, discussed earlier (above, p. 52), it may seem more than a coincidence that both texts appear to have omitted the preposition μετ᾽ from the phrase μετ᾽ εἰκάδας. If we can assume that some variation in the

---

[32] An examination of the stone in Athens in 1958 revealed that the letter in line 37 after Μαιμακτηριῶνος is epsilon, quite clear; but there is no letter preserved in the following space.

wording was permissible before the phraseology became stereotyped all epigraphical difficulties might be removed by reading, for example, δε[υτέραι εἰκάδων] in *I.G.*, II² 504, and ε[ἰκάδι ἕκτηι] in *I.G.*, II², 448. I have not introduced these readings into the texts, but I should like to emphasize how little we know of the history of the phrase μετ' εἰκάδας, of why it was introduced, and of how and when it came to have, in general, backward count.[33] Apparently the count was backward in *I.G.*, II², 448, line 37, while it was forward in *I.G.*, II², 504, lines 5-6.[34]

Designations of days of a month from the 20th to the end are now to be arranged as follows, when the count was made with μετ' εἰκάδας:[35]

### DAYS FROM THE 20TH TO THE 30TH OF THE MONTH IN CHANCERY STYLE FROM THE END OF THE FOURTH CENTURY B.C.

|  | Backward Count | | Forward Count |  |
|---|---|---|---|---|
| 20. | δεκάτη προτέρα | | δεκάτη προτέρα | 20. |
| 21. | δεκάτη ὑστέρα | | δεκάτη ὑστέρα | 21. |
|  | Full | Hollow | Full and Hollow |  |
| 22. ἐνάτη μετ' εἰκάδας | | ὀγδόη μετ' εἰκάδας | δευτέρα μετ' εἰκάδας | 22. |
| 23. ὀγδόη μετ' εἰκάδας | | ἑβδόμη μετ' εἰκάδας | τρίτη μετ' εἰκάδας | 23. |
| 24. ἑβδόμη μετ' εἰκάδας | | ἕκτη μετ' εἰκάδας | τετρὰς μετ' εἰκάδας | 24. |
| 25. ἕκτη μετ' εἰκάδας | | πέμπτη μετ' εἰκάδας | πέμπτη μετ' εἰκάδας | 25. |
| 26. πέμπτη μετ' εἰκάδας | | τετρὰς μετ' εἰκάδας | ἕκτη μετ' εἰκάδας | 26. |
| 27. τετρὰς μετ' εἰκάδας | | τρίτη μετ' εἰκάδας | ἑβδόμη μετ' εἰκάδας | 27. |
| 28. τρίτη μετ' εἰκάδας | | δευτέρα μετ' εἰκάδας | ὀγδόη μετ' εἰκάδας | 28. |

[33] G. Klaffenbach, *Gnomon*, XXI, 1949, p. 135, in his review of Pritchett and Neugebauer, *Calendars*, asks: "Wie wollen die Verf. bei ihrer Stellungnahme den Wechsel der älteren Ausdrucksweise φθίνοντος zu der jüngeren μετ' εἰκάδας erklären, wenn die Zählweise die gleiche geblieben ist? Nur als eine sprachliche, dann sogar unzutreffende Variante des Ausdrucks?"

[34] Pritchett, *B.C.H.*, LXXXI, 1957, p. 282 note 1, cites an argument of G. Daux based upon an Eretrian inscription (*I.G.*, XII, 9, 207, suppl. p. 178) as confirming the Pritchett-Neugebauer conclusions about the count of days with μετ' εἰκάδας. What Daux argues (*R.E.G.*, LXIII, 1950, pp. 253-254) is that δεκάτη μετ' εἰκάδα in the Eretrian text is equivalent to δεκάτη φθίνοντος of good Attic date and hence that the count was backward. This proves nothing new. Backward count with μετ' εἰκάδας has been formally proved since the publication of my article in *Hesperia*, IV, 1935, pp. 525-561. Daux cannot comprehend how the Athenians had backward count and at the same time occasionally forward count: but this difficulty is subjective.

[35] For the first 19 days, see the table on p. 45, above.

| Full | Hollow | Full | Hollow | |
|---|---|---|---|---|
| 29. δευτέρα μετ' εἰκάδας | ἔνη καὶ νέα | ἐνάτη μετ' εἰκάδας | ἔνη καὶ νέα | 29. |
| 30. ἔνη καὶ νέα | | ἔνη καὶ νέα | | 30. |

This table, and that shown above (p. 45), are substantially the same as the table published in *Hesperia*, IV, 1935, p. 535.[36] When the count was with φθίνοντος we now know that the 21st day of a hollow month was ἐνάτη φθίνοντος. Presumably the use of δεκάτη προτέρα did not officially replace εἰκοστή for the 20th day until the latter part of the fourth century B.C. Probably at the same time δεκάτη ὑστέρα became the settled designation for the 21st.[37]

[36] Pritchett, *B.C.H.*, LXXXI, 1957, p. 283, implies that these different ways of counting days might all be merged into one if we assume enough tampering with the festival calendar. See above, especially p. 51 note 19.

[37] The restoration of [εἰκοστ]εῖ as the date by month in *I.G.*, II², 768, of the year 251/0 is probably not justified. As Pritchett and Neugebauer remark (*Calendars*, p. 82 note 10), the only suitable restorations are [ἐνδεκάτ]ει and [δωδεκάτ]ει, and in either case tampering with the festival calendar must be assumed. The matter is not absolutely certain, but the use of εἰκοστή at so late a date would be surprising.

# CHAPTER IV

## The Logistai Inscription

The suggestion was made earlier that Aristotle's statement about the lengths of prytanies in his day does not establish a rigid and inflexible rule from which there can be no deviation.[1] What he says is explicit enough, that the first four prytanies of the year had 36 days each and that the last six had 35 days each, and by and large this is unquestionably correct. One does not deny the validity of the rule when one matches it against the epigraphical and other literary evidence and finds that it does not always agree. Nor does one deny its validity because intercalary years have prytanies of 38 and 39 days, of which Aristotle says nothing, instead of 35 and 36, or because Aristotle makes no allowance for the occasional need to add to some year an extra day. Nevertheless, it is true that students of the calendar (I among them) have in the past diverged farther than was necessary from his norm. Now our latest investigators have decided not to diverge at all.

The danger is that, on the other side, we may now have struck a wrong balance in the value of the evidence, some of which I plan to examine again, for there is nothing in the language of Aristotle, explicit as it is, that compels us to take his statement as more than a general rule. I plan to look first into some of the indirect evidence, for the argument has been advanced that the conciliar year of the fifth century was demonstrably inflexible, with a fixed pattern of prytanies, and each year like every other year, a foreshadowing, as it were, of the regularity of the con-

---

[1] Above, pp. 13-15. The text of Aristotle is quoted above, p. 8 note 9. Pritchett and Neugebauer, *Calendars*, p. 34, give a bibliography of paraphrases and echoes from the lexicographers and scholiasts.

ciliar year of the fourth century and a corroborative proof of its rigidity as defined by Aristotle.

In 1928 I made a study of the Athenian calendar in the fifth century based mainly on an inscription containing the audited accounts of money borrowed by the Athenian State from 426 to 422 B.C.[2] I have since made some changes in the text (as restored) and in the interpretation of this inscription.[3] But the fact remains that the most significant portion of the text covers a span of four conciliar years, a total of 1464 days, and that these years are not coterminous with the years of the festival calendar. This is agreed, I believe, by all who have studied the text, even by those who otherwise have questioned my use of the document as evidence for a reconstruction of the calendar in the fifth century.[4] My study gave years of 366, 368, 365, and 365 days; the most recent study has given years all of uniform length, each year having 366 days. Indeed, it is fundamental to this study that the text is advance evidence of the kind of regularity in the conciliar year which Aristotle asserts a century later. It is a different type of conciliar year, but the inscription has been so restored that regularity can be achieved with each year made up of six prytanies of 37 days each followed by four prytanies of 36 days each, a precursor (with different figures) of the Aristotelian norm.

Let us consider this inscription again. First, there is a misprint in the text as given by Pritchett and Neugebauer which should be corrected. They have in lines 43-44 the readings and restorations:

---

[2] Meritt, *The Athenian Calendar* (Cambridge, Mass., 1928).
[3] *Athenian Financial Documents* (Ann Arbor, Mich., 1932), pp. 128-151; "Borrowings in the Archidamian War," *Cl. Quart.*, XL, 1946, pp. 60-64. See also O. Broneer, *Hesperia*, IV, 1935, pp. 158-159; A. Oguse, *B.C.H.*, LIX, 1935, pp. 416-420.
[4] Pritchett and Neugebauer, *Calendars*, pp. 94-108; Pritchett, *B.C.H.*, LXXXI, 1957, pp. 293-298.

[ΧΧΗΗΗ <sup>v</sup> τόκος τούτοις ἐγένετο] ΓΗΓΑΔΔΔΗΙ <sup>v</sup> τετάρτε
δόσις ἐπὶ τῆς Αἰαντ[ίδος πρυτ]ανεί[ας ὀγ<sup>vv</sup>]

[δόες πρυτανευόσες ℎέκτει καὶ] εἰκοστῆι τῆς πρυτανείας ℍ
τόκος τοῦτο[ις ἐγέν]ετο ΧΓΗΗ[Η <sup>vvvvv</sup>]

and then in their synoptic table they have claimed the 26th day
of the eighth prytany as the 85th day from the end of the year.[5]
If one posits 36 days in each of the last three prytanies, then the
85th day from the end of the year must be counted as the 24th
day of the eighth prytany, rather than the 26th.[6] This can be
done, by restoring the lines in question as follows:

[ΧΧΗΗΗ <sup>v</sup> τόκος τούτοις ἐγένετο] ΓΗΓΑΔΔΔΗΙ <sup>v</sup> τετάρτε
δόσις ἐπὶ τῆς Αἰαντ[ίδος πρυτ]ανεί[ας ὀγδό]

[ες πρυτανευόσες τετάρτει καὶ] εἰκοστῆι τῆς πρυτανείας ℍ
τόκος τοῦτο[ις ἐγέν]ετο ΧΓΗΗ[Η <sup>vvvvv</sup>].

Surely this is what was intended. If the restoration here is correct, my own synoptic table for the year 423/2 can be made to
conform[7] by giving 37 days to the seventh prytany (VII) and
36 days to the eighth (VIII). I do not regard the adjustment as
necessary, or even desirable, but only as possible: it brings
syllabic division to the end of line 43.

But the question remains how much one should rely on
syllabic division at the ends of the lines from 31 to 46 in determining the restorations. Quite definitely there is not syllabic
division, except by accident, from line 1 down through line
26, though we note that an uninscribed space was apparently
left at the end of line 17 after πρῶτος, evidently so that the next
line could begin with the complete word ἐγραμμάτευε.[8] This is

---

[5] Pritchett and Neugebauer, *Calendars*, pp. 102-104.
[6] I have not found this error noted in any of the reviews of Pritchett and
Neugebauer's book.
[7] *The Athenian Calendar*, p. 67.
[8] See Plate I at the end of *The Athenian Calendar*.

the nearest approach to syllabic division; the uninscribed spaces before and after the numeral at the end of line 10 are normal, just as they are with every item of interest introduced in this year by the words τόκος τούτον.

ΠΙΣΘΟΔΟ
ΤΕΣ. . . . ?
ΗΔΠͰͰΤΟ
ΥΤΑΝΕΥ
ΙΣΕΠΙΤΕΣ
ΣΤΟΥΤΟΙΣ
ΣΥΝΑΡΧΟΝ
ΑΙΧΣΥΝΑΡ
ΟΝΤΕΣΕΠΙ

Fɪɢ. 2.  Suggested ends of lines 28-36
of the Logistai Inscription

From line 27 to line 37 my text continues with regular stoichedon lines of 75 letters, just as above, while the text now proposed as a substitute for it exhibits ends of lines as shown in Figure 2. The assumption made is "that the engraver was inconsistent in his use of the final letter-space of each line, and, in addition, that syllabic division was observed in part of the document."[9] But if syllabic division was a desideratum, one is surprised to find that the word πρυτανευόσες is divided in lines 31 and 32 so that ΠΡΥΤΑΝΕΥ followed by two uninscribed spaces ends line 31, while the rest of the word ΟΣΕΣ begins line 32. The division ought to be πρυτανευό|σες. But this would not

---

[9] Pritchett and Neugebauer, *Calendars*, p. 102. I had earlier observed that, with the exceptions of lines 10 and 17, there were 75 letter-spaces in each line down to line 37, that from line 38 to line 75 the number of letter-spaces in a line varies from 74 to 75, and below line 75 that there were never more than 74 spaces. Cf. *The Athenian Calendar*, p. 29.

allow the date for the payment on the first day of the prytany, as is required by the new dispensation:

Pritchett and Neugebauer

[- - - - - - - - - - - - - - - - <sup>v</sup> τρίτε δ]όσις ἐπὶ τὲς
'Ερεχθείδος πρυτανείας hε[βδόμες πρυτανευ <sup>vv</sup>]
[όσες πρότει τὲς πρυτανείας Ⲕ̄ΤΧ]ḤΗ - - - - - - - - - - - -

Meritt

[- - - - - - - - - - - - - - - - <sup>v</sup> τρίτε δ]όσις ἐπὶ τὲς
'Ερεχθείδος πρυτανείας hε[βδόμες πρυτανευόσ]
[ες δευτέραι τὲς πρυτανείας Ⲕ̄ΤΧ]ḤΗ - - - - - - - - - - - -

My own restoration filled out every line here to 75 letters. We all agree that the payment must have been made 510 days before the end of the quadrennium, but the new reconstruction of the calendar is such that one cannot allow the year 423/2 to have only 365 days with 37 days in the seventh or eighth prytany of 424/3. Syllabic division does not solve the problem of these lines, and in the unjustifiable division of πρυτανευόσες between lines 31 and 32 as well as in the other anomalies at the ends of lines 28 and 30 I find a small epigraphical difficulty, but none the less one that is real, in the new calendar scheme.

This may seem a trifling matter; so I turn to a difficulty of a more serious nature. My account of the second payment of this year (lines 29-31) was that it was made on Prytany III 12 and was outstanding for 645 days.[10] The text as restored meets all the epigraphical requirements, and reads as follows:

- - - - - - - - - - - - - - - δευτέρα δ[όσις ἐπὶ τὲς . .<sup>5</sup>. . .]
[ .ίδος πρυτανείας τρίτης πρυταν]ευόσες δοδεκάτει τὲς πρυτανείας ΔΔΤΤΤ[ΧΧΧΧΗΗⲔ̄ τόκος τού]
[τοις ἐγένετο ΧΧΧⲔ̄Ⲕ̄Ⲕ̄ΗΙΙΙΙΙ <sup>v</sup> - - - - - - - - - - - - - - - -]

[10] See *The Athenian Calendar*, p. 71.

Interest upon the principal was reckoned thus for 645 days: [11]

| | | |
|---|---|---|
| 20 Tal. | in one day yields 4 Dr. | |
| | in 645 days yields | 2580 Dr. |
| 3 Tal. 3250 Dr. | in one day yields 4¼ obols | |
| | in 645 days yields 4¼ × 645 = 2741¼ obols = | 456 Dr. 5¼ obols |
| 1000 Dr. | in one day yields $^{1000}/_{1250}$ × ¼ obol | |
| | = ⅕ = ⅑ + $^{1}/_{12}$ obol (approximately) | |
| | in 645 days yields 125$^{5}/_{12}$ obols = | 20 Dr. $5^{5}/_{12}$ obols |
| 23 Tal. 4250 Dr. | in 645 days yields | 3057 Dr. 5 obols [12] |

The text which Pritchett and Neugebauer have offered for lines 29-31 reads as follows: [13]

– – – – – – – – – – – – – – – – – *δευτέρα δ[όσις ἐπὶ τε̂ς . . ⁵. . .]*
*[. . ⁵. . . πρυτανείας τρίτες πρυταν]ευόσες δοδεκάτει τε̂ς πρυ-*
*τανείας* 𐅉𐅉TTT[XXXX𐅅HHHΔ𐅃𐅂𐅂 *τό* ᵛ]
[*κος τούτον* XXX𐅄ΔΔ𐅃𐅂𐅂IIIII ᵛ – – – – – – – – – – – – – – – –]

According to their calendar the payment was outstanding for 647 days. If interest on the suggested amount of principal is computed algebraically the resultant figure is 3080.08 Dr., a sum larger by more than two drachmai than that which Pritchett and Neugebauer must restore, and have restored, in the text. The computation may also be made without the aid of algebraic methods and decimal notation:

| | | |
|---|---|---|
| 20 Tal. | in one day yields 4 Dr. | |
| | in 647 days yields | 2588 Dr. |
| 3 Tal. 4500 Dr. | in one day yields 4½ obols | |
| | in 647 days yields 4½ × 647 = 2911½ obols = | 485 Dr. 1½ obols |
| 317 Dr. | in one day yields $^{317}/_{1250}$ × ¼ obol | |
| | = ¼ × ¼ obol (approximately) = $^{1}/_{16}$ obol | |
| | in 647 days yields 40½ obols = | 6 Dr. 4½ obols |
| 23 Tal. 4817 Dr. | in 647 days yields | 3080 Dr. |

This figure is again more than two drachmai higher than that

[11] See the table of computed values in *The Athenian Calendar*, p. 32.
[12] To the nearest obol. [13] Pritchett and Neugebauer, *Calendars*, p. 99.

restored by Pritchett and Neugebauer, who are compelled by the other numerals in the text, in their complex interrelationship, to restore 3077 Dr. 5 obols. This error in the amount of interest is intolerable. I noted years ago that accuracy, as we understand the term, is not in every instance attainable and that the results of computation sometimes involve approximations,[14] but an error of this magnitude can only show that the amount of the principal or the number of days has been incorrectly estimated. There is no difficulty with the computation, for the only approximation is that of taking $^{31}7\!/_{1250}$ as the equivalent of $\frac{1}{4}$. Indeed, the assumption of this approximation favors the computation of Pritchett and Neugebauer rather than otherwise, for it is actually lower than it should be, and gives in the final summation the lowest possible amount of interest. The fact that even so the amount of interest must be fully two drachmai (or more) higher than the calendar demands merely shows that the proffered text cannot be correct, and that the calendar dates based upon it are of no value.

It is well to emphasize this particular discrepancy because Pritchett and Neugebauer have suggested some theoretical approximations, assuming, for example, that both $\frac{1}{5}$ and $\frac{1}{7}$ (which I had suggested should be avoided) should be approximated to $\frac{1}{6}$; and since interest on 1000 Dr. in one day amounts to $\frac{1}{5}$ obol and interest on $714\frac{2}{7}$ Dr. in one day amounts to $\frac{1}{7}$ obol, they draw the conclusion that a given interest of $\frac{1}{6}$ obol (using the reverse process of finding the principal) may represent any amount of principal between 1000 Dr. and $714\frac{2}{7}$ Dr. They illustrate with a table: [15]

---

[14] *The Athenian Calendar*, p. 37.
[15] Pritchett and Neugebauer, *Calendars*, p. 98.

| Principal | Accurately Computed Interest |
|---|---|
| 1250 drachmae | ¼ obol |
| 1000 " | ⅕ " |
| 833⅓ " | ⅙ " |
| 714²⁄₇ " | ⅐ " |

and then continue with this observation: "Actually, we think, one should go even further. Because ¼ obol is the smallest monetary unit which can appear in a result, we should face the possibility that amounts up to 1250 drachmae in the principal may escape us."

This is deceptive, and leads to error. The suggestions are based, obviously, on interest for one day only, and they presuppose the reckoning of principal from interest, which is not our problem. Nor was it an Athenian's problem. The correct procedure to follow, when we are making restorations in this text and have the interest known but not the principal, is to free ourselves from all approximations and reckon the principal algebraically with as much accuracy as possible. Then we must test the amount of principal so determined and test also amounts slightly larger and slightly smaller, such as can be restored epigraphically in the text, to see whether the application of any known breakdown and its approximations, or any known Athenian method of reckoning, will yield, in the time allotted, the amount of interest from which the reckoning started. Clearly this test, which is essential, was not applied to the restorations which Pritchett and Neugebauer have made in lines 30-31. Nor can I find that any reviewer of their book, many of whom have accepted their arithmetical offerings, has discovered and exposed this fallacy.[16]

---

[16] The claim of Pritchett and Neugebauer (*Calendars*, p. 100) that "the maximum error permitted in any given method of computing interest should likewise be allowed in computing the principal" is to be rejected. Once a restoration has been made which involves principal and interest, it is obvious

Let us return for a moment to trivial matters, for they too have an epigraphical interest. Our latest investigators have proposed for lines 21-22 the following change of restoration: [17]

[————ἐπὶ] τῆς Πανδιονίδος πρυτανείας ἐνάτες πρυτ[ανενόσες ὀγδό ᵛ]
[ει καὶ] δεκάτει ἐμέραι τῆς π[ρυταν]είας ἐσελελυθυίας Ⱶ τόκος τούτοις ἐγένε[το ΤΤΧΧΧ𐅄ΗΗΔΔΔ].

The date in line 21 is too short by one letter-space to fill out the full line of 75 letters, the normal number without syllabic division both before and after, while it is assumed that the twelve figures of the amount of interest in line 22 "were inscribed in the final 11 letter-spaces," condensing a line that really would have been too long by one letter for the space on the stone. Had the engraver wished to get a numeral of twelve spaces on the stone at the end of line 22, he had only to treat the broad figure Ⱶ earlier in the line just as he in fact treated the broad figure Ⱶ in line 44, and the line could have remained stoichedon to the end.[18] His lack of anxiety about the extra space taken up by the figure Ⱶ in line 22 is some indication that the numeral at the end of the line was not overly long.[19]

But the vital matter is the failure of the new restorations in the third year. This means their failure also in all other years of the quadrennium, and, incidentally, we return here to the shorter numeral ΤΤΧΧΧ𐅄ΗΗ𐅂Δ for the amount of interest at the

that the only test of its validity is the reckoning from principal to interest, not *vice versa*.

[17] Pritchett and Neugebauer, *Calendars*, p. 102.
[18] A facsimile drawing may be consulted in Meritt, *The Athenian Calendar*, Plate I.
[19] I record my dissent from the recent treatment of the ends of these, and other, lines, especially 21, 22, 30, 31, 38, 40, 41, and 42, because in support of it Pritchett has recently noted that "several reviewers have regarded our text as equally judicious, epigraphically speaking, as that of any of our predecessors" (*B.C.H.*, LXXXI, 1957, p. 292), with especial reference to A. M. Woodward's review in *J.H.S.*, LXVIII, 1948, p. 166.

end of line 22.[20] We must, indeed, reject the whole assumption of regularity in the conciliar year. The quadrennium from 426 to 422 contained 1464 days, but the four years were not divided out evenly with 366 days in each. The first year had 366 days, the second 368 days, the third 365 days, and the fourth 365 days. So far as "regularity" in the calendar of the fifth century is concerned, we have the same knowledge that we had in 1928: the prytanies contained regularly, during these four years, 36 or 37 days, but they show no definite order of precedence or alternation.[21]

Were I now writing the study that I made in 1928 I should develop two of the computations of principal and interest differently:

### I. FIRST PAYMENT OF THE THIRD YEAR (LINE 29)

| | | |
|---|---|---|
| 30 Tal. | in one day yields 6 Dr. | |
| | in 705 days yields | 4230 Dr. |
| 2 Tal. 5500 Dr. | in one day yields 3½ obols | |
| | in 705 days yields 3½ × 705 = 2467½ obols = 411 Dr. 1½ obols | |
| 1050 Dr. | in one day yields $^{1050}/_{1250}$ × ¼ obol | |
| | = $^{5}/_{24}$ obol (approximately) = ⅛ + $^{1}/_{24}$ obol | |
| | in 705 days yields 117½ + 30 = 147½ obols = 24 Dr. 3½ obols | |
| 33 Tal. 550 Dr. | in 705 days yields | 4665 Dr. 5 obols |

### II. LAST PAYMENT OF THE FOURTH YEAR (LINE 46)

| | | |
|---|---|---|
| 15 Tal. | in one day yields 3 Dr. | |
| | in 34 days yields | 102 Dr. |
| 2 Tal. 5500 Dr. | in one day yields 3½ obols | |
| | in 34 days yields 3½ × 34 = 119 obols = | 19 Dr. 5 obols |
| 500 Dr. | in one day yields $^{500}/_{1250}$ × ¼ obol. | |
| | = ⅛ obol (approximately) | |
| | in 34 days yields | 3½ obols |
| 18 Tal. | in 34 days yields | 122 Dr. 2½ obols |

I expressed my belief earlier that the concluding figures of the principal, as I then thought it had to be restored in line 46 (ΔⲂTTTHΔΔⲄⲄIIᏟ), were erroneously added to the true

---

[20] Meritt, *Athenian Financial Documents*, p. 138.
[21] Meritt, *The Athenian Calendar*, p. 71.

principal (ΔⲒ̄ΤΤΤ) because the eye of the scribe, when he copied, had been misled by the amount of the interest (ΗΔΔⲒⲒΙⲤ).[22] The total of the principal in line 47 demands the conflated figure, but it may not after all have been inscribed in line 46 and certainly the conflating increment of ΗΔΔⲒⲒΙⲤ was not used in the reckoning of interest. I should now write line 46 as follows:

[ας ΔⲒ̄ΤΤΤ τόκος τούτοι ἐγένετο] ΗΔΔⲒⲒΙⲤ ᵛ κεφάλαιον
τὸ ἀρχαίο ἀναλό[ματος] ἐπὶ τὲς Τι[μοκλέο]

with a note that the interest as well as the principal got taken into the total of the principal of the year in line 47. There is not room for the usual phrase τόκος τούτοις ἐγένετο; hence I have suggested the singular (as it appears throughout the payments of the Other Gods in lines 60-97). Alternatives are to crowd in an extra letter in writing the normal plural (which I dislike to do), or perhaps to write ΔⲒ̄ΤΤΤ ⟨τ⟩όκος τούτοις ἐγένετο or ΔⲒ̄ΤΤΤ τόκος τούτον ἐγένετο.[23]

Knowledge that the conciliar year of the fifth century did not conform to any set or regular pattern takes the evidence, so-called, of regularity in the fifth century out of the catalogue of proof that regularity existed in the fourth century. The argument has run somewhat as follows: (1) Aristotle's rule of prytanies in the late fourth century must be rigidly followed; (2) the same rigidity was probably observed in the conciliar year in other centuries; (3) the epigraphical record of the borrowings of the Athenian State from 426 to 422 B.C. shows a rigid calendar in which each year had 366 days made up of

---

[22] Meritt, The Athenian Calendar, p. 66.

[23] Throughout the text from line 1 to line 52 the phrase τόκος τούτον is regularly followed by an uninscribed space, the phrase τόκος τούτοις ἐγένετο never. This is one reason for rejecting my earlier restoration τόκος τούτον in line 46, and an additional reason for rejecting the restoration of Pritchett and Neugebauer in line 31 (cf. Calendars, p. 99).

ten prytanies the first six of which had 37 days each and the last four of which had 36 days each; hence (4) the scheme of the fifth century (with no flexibility) is an analogical argument that supports the interpretation of Aristotle's text to mean a rigid conciliar year also in the fourth century. But in fact the supposed rigidity of the fifth century did not exist. So by change and change about we argue that the flexibility of the fifth century gives some reason to expect flexibility also in the fourth.

We must now turn our attention to the fourth and later centuries. There are still a number of improvements that can be made in the epigraphical record between 341 and 307 B.C., and we shall find that in this as well as in the later centuries no regular rule of prytanies can be applied.

# CHAPTER V

## THE FOURTH CENTURY

The conciliar year may follow Aristotle's definition of it exactly. It would be strange if in general it did not, for Aristotle was enunciating a general rule; but I assume variation of prytanies of 36 and 35 days in an ordinary year, holding even that at times a prytany of 37 days may be possible, as is shown by the first prytany of 407/6,[1] and possibly also by the sixth prytany of 408/7.[2] In intercalary years there were prytanies of 38 and 39 days, but their position too within the year might vary, as we have already found to be the case in 336/5 B.C. as studied above (pp. 10-15). I follow here the plan adopted by Pritchett and Neugebauer in giving the evidence for individual years, with notes on the interrelation of the festival and conciliar calendars.[3]

### 346/5 (Ordinary)
### Archon Archias

The earliest calendar equation preserved from the fourth century is from an inscription of the Athenian colony on Samos:[4]

---

[1] See below, pp. 213-214.
[2] But see below, pp. 214-215.
[3] *Calendars*, pp. 34-67.
[4] Carl Curtius, *Inschriften und Studien zur Geschichte von Samos* (Progr. Lübeck, 1877), p. 11, lines 56-57; cf. Pritchett and Neugebauer, *Calendars*, p. 41. Peisileos was archon on Samos when Archias was archon at Athens. One should note that this text is not formal evidence for the calendar at Athens. The dates are dates recorded by the Athenians on Samos, and, though conformity to Athens in such larger matters as intercalation may be expected, local distribution of months and prytanies may well have differed.

*a.* 346/5 *a.*

Ἐπὶ Πεισίλεω ἄρχοντος, μηνὸς Ποσιδειῶνος τετράδι φθίνοντος,
ἐπὶ τῆς
Πανδιονίδος πέμπτης πρυτανείας μιᾶι καὶ τριακοστεῖ, — — — —

Posideon 27 = Prytany V 31 = 175th day.

The equation is that suggested by Pritchett and Neugebauer,
and if correct then the first four prytanies had each 36 days.
But it is equally legitimate to take Ποσιδειῶνος τετρὰς φθίνοντος
as the 26th day of a hollow month,[5] thus making the equation
read

Posideon 26 = Prytany V 31 = 174th day.

There is in either case no irregularity in the calendar, though
the 174th day would require that only three of the first four
prytanies have 36 days each.

341/0 (Intercalary)
Archon Nikomachos

*I.G.*, II², 229 and *Addenda* (p. 659); above, p. 10

*a.* 341/0 *a.* ΣΤΟΙΧ. 38

[θ        ε]        ο        [ι]
['Επὶ Νικομάχου ἄρχοντος ἐπ]ὶ τῆς Λεων[τίδ]ο[ς δεκ]
[άτης πρυτανείας ἧι 'Ονήσι]ππος Σμικύ[θο] 'Α[ραφήν]
[ιος ἐγραμμάτευεν· ἔνηι κα]ὶ [ν]έαι, ἑβδ[όμηι] κ[αὶ τρ]
5    [ιακοστῆι τῆς πρυτανεία]ς· — — — — — κτλ. — — — — — —

338/7 (Ordinary or Intercalary)
Archon Chairondes

*I.G.*, II², 237 and *Addenda* (p. 659) [6]

---

[5] See the table above, p. 45.
[6] For the secretary of this year, see G. A. Stamires, *Hesperia*, XXVI, 1957,
p. 243.

74            The Fourth Century
a. 338/7 a.                                    ΣΤΟΙΧ. 41

['Επὶ Χαιρών]δου ἄρχο[ντος ἐπὶ τῆς Πανδιονίδος δ]ε[κ]ά
[της] π[ρυτα]νείας ἧι Φ[ίλιππος 'Αντιφήμου Εἰρεσίδης]
[ἐγρ]αμ[μά]τευεν· Θαργη[λιῶνος τετράδι φθίνοντος, τρ]
[ίτη]ι [τῆ]ς πρυτανείας· — — — — — κτλ. — — — — — — — —

The equation here given, and accepted by Kirchner (in the
*Addenda*) and by Pritchett and Neugebauer, was suggested by
B. Haussoullier (*Rev. crit.*, XLVII, 1899, p. 406), who attributed
the days of the year to their several prytanies according to
Aristotle's norm (4 × 36 + 6 × 35). The equation does, in fact,
conform to that norm, if Thargelion was full and Skirophorion
hollow. The equation proposed by Adam Reusch, and printed
by Kirchner in *I.G.*, II², 237, exhibits a different text: [7]

['Επὶ Χαιρών]δου ἄρχο[ντος ἐπὶ τῆς Πανδιονίδος δ]ε[κ]ά
[της] π[ρυτα]νείας ἧι Φ[ίλιππος 'Αντιφήμου Εἰρεσίδης]
[ἐγρ]αμ[μά]τευεν· Θαργη[λιῶνος δευτέραι φθίνοντος, ἕ]
[κτη]ι [τῆ]ς πρυτανείας· — — — — — κτλ. — — — — — — — —

Reusch had experimented with a number of restorations, including some with μετ' εἰκάδας instead of φθίνοντος, but he finally
hit upon that recorded here "quia verum diem invenisse me ex
Aeschine spero. Ibi enim III 27 legitur: ἐπὶ γὰρ Χαιρώνδου
ἄρχοντος Θαργηλιῶνος δευτέρᾳ φθίνοντος ἐκκλησίας οὔσης ἔγραψε
Δημοσθένης ἀγορὰν ποιῆσαι τῶν φυλῶν Σκιροφοριῶνος δευτέρᾳ
ἱσταμένου καὶ τρίτῃ." [8] Reusch was writing before the discovery
of Aristotle's 'Αθηναίων Πολιτεία, and his belief that this decree
was passed on the same day that Demosthenes spoke in the
assembly was not colored by Aristotle's "rule," though that rule
was already indirectly known from several scholia.

[7] For the secretary, see G. A. Stamires, *Hesperia*, XXVI, 1957, p. 243.
[8] Adam Reusch, *De diebus contionum ordinariarum* (Diss. Strassburg, 1879),
p. 8. This was republished in part in *Dissertationes Philologicae Argentora-
tenses*, III (Strassburg, 1880).

So we have the good evidence that there was a meeting of the Ekklesia on the next to last day of Thargelion, and a restored equation in an epigraphical text which gives either the same day or the second preceding day. Our choice should, I think, favor Reusch's original proposal, for we should not expect two meetings of the Ekklesia separated by only one day.

Meetings of the Ekklesia are known in which the business ran over into a second day, or was resumed on a second day. Such, for example, are the two consecutive days of the Mytilenaean debate (Thuc., III, 36, 4-6), the meetings in the crisis over Philip of Macedon in 346 B.C. (Aischines, II, 61-63; III, 68), and, epigraphically, the meetings from which we have the two decrees *I.G.*, II², 504 and 505 (above, p. 52). The possibility was envisaged also in 425 B.C. for the discussion of tribute.[9] Other examples could be cited. But these meetings can all be thought of as prolongations of one Ekklesia. Usually, between meetings of the Ekklesia, the Council had to exercise its probouleutic function and prepare the new agenda. This required time, and would have been preliminary routine for the summoning of a second Ekklesia.[10]

I regard the restoration Θαργη[λιῶνος τετράδι φθίνοντος, τρίτη]ι [τῆ]ς πρυτανείας, therefore, as very unlikely, and prefer that which puts the meeting on δευτέρα φθίνοντος when it is known from Aischines that a meeting was held. The consequence that Prytany X must in this case have had 36 days seems to me to violate the letter, perhaps, but not the spirit of Aristotle's rule.

The alternative is to separate the meetings by a wider interval of days. This can be done, I think, on the assumption of an intercalary year with the equation Θαργη[λιῶνος ἑβδόμηι φθίνον-

---

[9] See Meritt, Wade-Gery, McGregor, *Athenian Tribute Lists*, II, p. 41 (A9, lines 33-36).
[10] For the formalities, see Busolt-Swoboda, *Gr. Staatskunde* (1926), p. 993.

τος, τρίτη]ι [τῆ]ς πρυτανείας. If Skirophorion was hollow Prytany X had 38 days; if Skirophorion was full Prytany X had 39 days. Or the equation might be restored as Θαργη[λιῶνος δεκάτηι φθίνοντος, πρώτη]ι [τῆ]ς πρυτανείας, with Skirophorion hollow and the last prytany having 39 days. There is no sure evidence for the calendar character of the year 338/7.

337/6 (Ordinary)

Archon Phrynichos

*Hesperia*, IX, 1940, pp. 325-327 (35)

a. 337/6 a.                                    ΣΤΟΙΧ. 39

['Επὶ Φ]ρυνίχο[υ ἄρχοντος ἐπὶ τῆς 'Ακαμαντίδος ἕκτ]
[ης π]ρυταν̣ε[ίας ἦι Χαιρέστρατος 'Αμεινίου 'Αχαρν]
[εὐ]ς̣ ἐγραμμ[άτευεν· Γαμηλιῶνος ἑβδόμηι ἱσταμέν]
[ου], πέμπτ[ηι τῆς πρυτανείας· τῶν προέδρων – – κτλ. –]

*I.G.*, II², 239 (cf. Schweigert, *Hesperia*, IX, 1940, p. 327)

[θ]            ε            [ο            ι]
                    'Αλκίμα[χος]

                                    ΣΤΟΙΧ. 26

['Ε]πὶ Φρυνί[χου ἄρχοντος ἐπὶ τῆς 'Α]
[κ]αμαντίδ[ος ἕκτης πρυτανείας ἦ]
5   [ι] Χαιρέσ[τρατος 'Αμεινίου 'Αχαρν]
[εὐ]ς ἐγρα[μμάτευεν· Γαμηλιῶνος ἑ]
[βδόμ]ηι [ἱσταμένου, πέμπτηι τῆς π]
[ρυτανείας – – – – – κτλ. – – – – –]

These two texts complement each other, and embody decrees passed on the same day, yielding the equation:

[Gamelion 7] = Prytany [VI] 5 = 184th day.

With regularly alternating full and hollow months the seventh of Gamelion was the 184th day of the year; the first four prytanies each had 36 days.

*Hesperia*, XXI, 1952, pp. 355-356 (5)

The text shows that Leontis was the ninth prytany, but there is no calendar equation.

*Hesperia*, VII, 1938, pp. 292-294 (19)

*a.* 337/6 *a.* ΣΤΟΙΧ. 40

['Επὶ Φρυνίχου ἄρχοντος ἐπὶ τῆς Πανδιονίδος δ]εκά
[της πρυτανείας· Χαιρέστρατος 'Αχαρνεὺς ἐγρ]αμμά
[τευεν· Σκιροφοριῶνος ἕκτηι ἐπὶ δέκα, δευτέρ]αι κ[α]
[ὶ εἰκοστῆι τῆς πρυτανείας· τῶν προέδρων – –] – κτλ. –

[Skirophorion 16] = Prytany X [22] = 341st day.

*I.G.*, II², 242

*a.* 337/6 *a.* ΣΤΟΙΧ. 33

[θ] ε [ο ί]
['Ε]πὶ Φρυνίχ[ο]υ [ἄρχοντος ἐπὶ τῆς Πανδιονί]
[δ]ος δεκάτ[ης πρυτανείας· Χαιρέστρατος 'Α]
χαρνεὺς ἐγ[ραμμάτευεν· Σκιροφοριῶνος ἕ]
5 νει [κ]αὶ νέ[αι, πέμπτηι καὶ τριακοστῆι τῆς]
πρυτανεί[ας – – – – – κτλ. – – – – – – – – –]

[Skirophorion] 29 = Prytany X [35] = 354th day.

*I.G.*, II², 243, was passed on the same day as *I.G.*, II², 242, but it has no calendar dates. *I.G.*, II², 240 and 241, were also passed during the tenth prytany, on one day, but on a day different from that of *I.G.*, II², 242 and 243. They have no calendar equations.

I suggest that the year was ordinary, with 355 days, and that the months and prytanies were arranged as follows:

Months 30 29 30 29 30 29 30 29 30 29 30 30 = 355
Prytanies 36 36 36 36 35 35 35 35 35 36 = 355

There is no irregularity. I assume here a year of 355 days because such is the length suggested by the lunar tables published by Parker and Dubberstein.[11] But I hold this to be by no means an inevitable guide, and consider a year of 354 days entirely possible. For the suggestion that Σκιροφοριῶνος ἔνη καὶ νέα may have been followed by ἔνη καὶ νέα ἐμβόλιμος, see above, pp. 10, 12-13.

### 336/5 (Intercalary)
### Archon Pythodelos [12]

*I.G.*, II², 328

a. 336/5 a.                                  ΣTOIX. 28

['Επὶ Πυθοδήλου ἄρχοντος ἐπ]ὶ τῆς 'A[κ]
[αμαντίδος τετάρτης πρυτ]ανείας ἧ
[ι . . . . . . . . . . . . . . .¹⁹ . . . . . . . . ἐ]γραμμάτ
[ευεν· Μαιμακτηριῶνος τετ]ράδι φθί
5 [νοντος, ὀγδόει καὶ εἰκοστ]εῖ τῆς πρ
[υτανείας· τῶν προέδρων − −] − κτλ. −

[Maimakterion] 26 = Prytany [IV 28] = 144th day.

*I.G.*, II², 330 (lines 29-30)

a. 336/5 a.                                  ΣTOIX. 48

['Επ]ὶ Πυθοδήλου ἄρχοντος [ἐπὶ τῆς ... ηίδος ἐνάτης πρυτανεί]
[ας], τετράδι ἐπὶ δέκα, δευτ[έραι τῆς πρυτανείας· − − κτλ. − ]

⟨Mounichion⟩ 14 = Prytany [IX] 2 = 309th day.

---

[11] *Babylonian Chronology*, p. 35.
[12] See above, pp. 10-15.

*I.G.*, II², 330 (lines 47-49)

*a.* 336/5 *a.*                                          ΣΤΟΙΧ. 48

['Επ]ὶ Πυθοδήλου ἄρχοντος ἐπ[ὶ τῆς ...ᶜ... ίδος δεκάτης
                                                            πρυτα]
[ν]είας, ἔνει καὶ νέαι, ἑβδόμη[ι καὶ τριακοστῆι τῆς πρυτανεία]
[s]· ἐκκλησία· — — — — — — — — — — κτλ. — — — — — — — — — — —

⟨Skirophorion⟩ 29 = Prytany [X] [3]7 = 383rd day.

The procession of months and prytanies throughout the year
may be tabulated as follows:

Months   29  30  29  30  29  30  29  30  29  30  29  30  30 = 384
Prytanies 39  39  38  38  38  38  38  39  39  38 = 384

335/4 (Ordinary)
Archon Euainetos

*I.G.*, II², 330 (lines 1-3) [13]

*a.* 335/4 *a.*                                          ΣΤΟΙΧ. 46

['Επὶ Ε]ὐαινέτου ἄρχο[ντος ἐπὶ τῆς ...ᶜ... ίδος τρίτης πρυτ]
[ανε]ίας, ἧι Πρόξενος Π[υλαγόρου 'Αχερδούσιος ἐγραμμάτευ]
[εν]· ἔνηι καὶ νέαι, ἑβδό[μηι καὶ δεκάτηι τῆς πρυτανείας· – κτλ.]

⟨Boedromion⟩ 30 = Prytany [III] [1]7 = 89th day.

The name of the month was not inscribed, but must have
been Boedromion. If the first three months had 29, 30, and 30
days, and if the first two prytanies had 36 days each, the calendar equation is regular.

*Hesperia*, IX, 1940, pp. 327-328 (36)

*a.* 335/4 *a.*                                          ΣΤΟΙΧ. 26

['Επὶ Εὐα]ινέτου ἄρχοντος ἐπὶ τῆ[s]

---

[13] The name of the phyle was restored in the *Corpus* as Antiochis, now
known to have held the tenth prytany (see below).

['Ακαμαν]τίδος πέμπτης πρυτανε[ί]
[ας ἦι Πρόξ]ενος Π[υλα]γόρου 'Αχερ[δ]
[ούσιος ἐ]γραμμάτ[ευ]εν· Ποσειδε[ῶ]
5 [νος ἐνδεκ]άτει· — — — — κτλ. — — — — —

The number of the day within the prytany was not inscribed, but in an ordinary year the eleventh day of Posideon (or twelfth, reading [δωδεκ]άτει) must have fallen well within the fifth prytany.

*I.G.*, II², 333

a. 335/4 a.

['Επὶ Εὐαινέτου ἄρχοντος ἐπὶ τῆς 'Αντιοχίδος δεκά]της,
Σκιροφορ[ιῶνος ἕκ]τηι ἱσταμένου· — — — — κτλ. — — — — —

The number of the phyle is known from two other texts of this year (see below). The number of the day within the prytany was not inscribed, but the sixth of Skirophorion must have fallen normally well within the tenth prytany.

E.M. 13067 [14]

a. 335/4 a.                ΣΤΟΙΧ. 25

[θ]          ε          ο          [ί]
['Επ' Εὐαιν]έτου ἄρχοντος ἐπὶ [τῆς]
['Αντι]οχίδος δεκάτης πρυτα[νεί]
[ας ἦι] Πρόξενος Πυλαγόρου ['Αχερ]
5 [δούσ]ιος ἐγραμμάτευεν· Σ[κιροφ]
[ορι]ῶνος ὀγδόηι ἐπὶ δέκα, [τ]ρ[ίτη]
[ι κ]αὶ εἰκοστῆι τῆς πρυταν[είας·]
κτλ.

Skirophorion 18 = Prytany X 23 = 343rd day.

---

[14] G. A. Stamires has kindly supplied to me the text of the opening lines of this unpublished inscription.

*I.G.*, II², 331

a. 335/4 a.                              NON-ΣΤΟΙΧ.

['Επ' Εὐαινέτου ἄρχοντος ἐπὶ τῆς 'Αντιο]χίδος
[δεκάτης πρυτανείας ἧι Πρόξενος Πυ]λαγόρο
['Αχερδούσιος ἐγραμμάτευεν· ἔνηι] καὶ νέα[ι,]
[πέμπτηι καὶ τριακοστῆι τῆς πρυταν]είας·
5 [ἐκκλησία· – – – – – – – κτλ. – – – – – – –

⟨Skirophorion⟩ 30 = Prytany [X 35] = 355th day.

The restorations are new, based on the evidence of E.M. 13067
(just above). Since the phyle Antiochis was in prytany, the
month was either Thargelion or Skirophorion, and of these Thargelion may be eliminated because no equation with the prytany is
possible for Thargelion 29 or 30. The lettering is not stoichedon, or, rather, not strictly stoichedon; [15] so one cannot state
categorically that the restoration τετάρτηι καὶ τριακοστῆι in line 4
is impossible. But the restoration πέμπτηι καὶ τριακοστῆι gives
a one-to-one correspondence of letters (so far as preserved) with
line 3, and eases, rather than exacerbates, the problem of the
calendar.

If one assumes a final prytany of 35 days and a final month
of 30 days, he may hold that Hekatombaion and Metageitnion
counted 59 days and that beginning with Boedromion there
was an alternation of months, once reversed, down through
Skirophorion. This involves no irregularity in the calendar, for
it was a long recognized necessity that occasionally two full
months should come together. The year will now have had
355 days, and Aristotle's norm of four 36-day prytanies followed by six 35-day prytanies will be modified to the extent
of having four 36-day prytanies followed by five 35-day pry-

[15] Line 1 had 35 letters; line 2 had 35 letters of which the last three occupied
two spaces; line 3 had 34 letters, completely filling the line; and line 4 had
33 letters. The patronymic in line 2 was spelled Πυλαγόρο.

tanies and one 36-day prytany (not, however, the tenth). Pritchett and Neugebauer allow variation in length of prytany in such a case only in the tenth prytany at the end of the year. But since this is merely a hypothesis, and since Aristotle says nothing about the contingency,[16] one may assume that the extra day was added to some prytany earlier than the tenth, closer perhaps in date to the time when the second (consecutive) 30-day month was introduced. We thus have a normal ordinary year of 355 days, normal, that is, in both calendars, in which the months and prytanies were arranged as follows:

Months   29   30   30   29   30   29   30   30   29   30   29   30 = 355
Prytanies   36   36   36   36   35   35   36   35   35   35 = 355

Pritchett and Neugebauer have removed from this year the fragmentary text *I.G.*, II², 332. I agree with them that it proves nothing about the calendar.[17]

334/3 (Ordinary)

Archon Ktesikles

*I.G.*, II², 336[18]

*a.* 334/3 *a.*

['Επὶ Κτησικλέους ἄρχοντος ἐπὶ τῆς]
['Ιπποθωντίδος τετάρτης πρυτανείας]
ἧι Μνησίφιλος Μνήσω[νος Φαλη ἐγραμμά]    ΣΤΟΙΧ. 31
τευεν· Μαιμακτηριῶν[ος ἐνδεκάτηι, μιᾶ]
5   ι καὶ εἰκοστεῖ τῆς π[ρυτανείας· ἐκκλησ]
ία κυρία· – – – – – – – κτλ. – – – – – –

Maimakterion [11] = Prytany [IV] 2[1] = 129th day.

---

[16] Pritchett and Neugebauer, *Calendars*, p. 36. My basic assumption (and interpretation of Aristotle) is again more flexible.
[17] *Calendars*, p. 45.
[18] For the demotic of the secretary, see Schweigert, *Hesperia*, IX, 1940, p. 340.

If the first four months had 118 days, then Maimakterion 11 was the 129th day, and the first three prytanies had 108 days, or a normal 36 days each. There are no irregularities. Presumably the year began and ended with hollow months, with one alternation reversed at some time after Posideon:

Months  29  30  29  30  29  30  29  30  30  29  30  29 = 354
Prytanies  36  36  36  36  35  35  35  35  35  35 = 354

I.G., II², 335 (cf. Hesperia, IX, 1940, pp. 339-340)

a. 334/3 a.                ΣΤΟΙΧ. 21

```
   [θε]οἱ
   ['E]πὶ Κτησικ[λέους ἄρχοντο]
   [ς] ἐπὶ τῆς 'Ακα[μαντίδος ἐνά]
   [τ]ης πρυτανε[ίας ἧι Μνησίφ]
5  [ι]λος Μνήσ[ω]ν[ος Φαληρεὺς ἐ]
   [γ]ραμμάτευε[ν· Μουνιχιῶνο]
   [ς] ἕκτηι μετ' [εἰκάδας· ἐκκλη]
   [σ]ία· – – – – – κτλ. – – – – – –
```

There is here no equation, for the date within the prytany is missing. Any date within the last decade of Mounichion would have fallen in the ninth prytany. The same calendar formula has been restored in I.G., II², 405 and 414a.[19]

333/2  (Intercalary)
Archon Nikokrates

I.G., II², 337
a. 333/2 a.        ΣΤΟΙΧ. 20

```
[θ]      ε      ο      ι
'Επὶ Νικοκράτους ἄρχοντ
ος ἐπὶ τῆς Αἰγεῖδος πρώτ
ης πρυτανείας· τῶν – κτλ. –
```

[19] See Eugene Schweigert, Hesperia, IX, 1940, p. 340.

There is no equation, but the text is evidence for the name of the first prytany.

*I.G.*, II², 338 [20]

Metageitnion 9 = Prytany I 39 = 39th day.

*I.G.*, II², 339 [21]

Metageitnion [2]4 = Prytany II 1[5] = 54th day.

*I.G.*, II², 340

a. 333/2 a.　　　　　ΣΤΟΙΧ. 23

['Επὶ Νικοκράτους ἄρχοντος ἐ]
[πὶ τῆς . . .⁵. . . ίδος τετάρτης π]
[ρυτανείας ἧι 'Αρχέ]λας Χαι[ρί]
[ου Παλληνεὺς ἐγρ]αμμάτευ[εν·]
5　[Μαιμακτηριῶνος] ἐνδεκάτ[ηι,]
[δωδεκάτηι τῆς πρ]υτανεία[ς· ἐ]
[κκλησία κυρία· τῶν] προέδρ[ων]
κτλ.

[Maimakterion] 11 = Prytany [IV 12] = 129th day.

With Prytanies I, II, and III having 39 days each, the 12th day of the fourth prytany is the 129th day of the year. This is also the 11th day of Maimakterion, if the first four months had 30, 29, 30, 29 days.

*I.G.*, II², 358 (cf. Dinsmoor, *Archons*, p. 357)

a. 333/2 a.

['Επὶ Νικοκράτου]ς ἄρχο[ντος]
[ἐπὶ τῆς . . . . . . . . . . . .¹⁴. . . . . . .]της πρυτανείας ἧι　ΣΤΟΙΧ. 35
['Αρχέλας Χαιρίου Παλληνε]ὺς ἐγραμμάτευεν· 'Ε

---

[20] For the text and discussion, see above, pp. 48-51.
[21] See above, pp. 48-51.

[λαφηβολιῶνος ἕνηι καὶ ν]έαι ἐμβολίμωι, πέμ
5 [πτηι καὶ εἰκοστῆι τῆς πρ]υτανείας· – – κτλ. –

Elaphebolion 30 = Prytany {VIII} [2] 5 = 296th day.

Stoichedon order could be preserved in line 3 by writing the patronymic as Χαιρίο (cf. [Πυ]λαγόρο in *I.G.*, II², 331 of the year 335/4). In spite of the letters preserved in line 2, the number of the prytany was surely eighth.[22] I suggest that the pattern of the year be restored with months and prytanies arranged as follows:

|        |    |    |    |    |    |    |    |    | (29+1) | (30-1) | (29) | (30) |        |
|--------|----|----|----|----|----|----|----|----|--------|--------|------|------|--------|
| Months | 30 | 29 | 30 | 29 | 30 | 29 | 30 | 29 | 30     | 30     | 29   | 30   | 29 = 384 |
| Prytanies | 39 | 39 | 39 | 39 | 39 | 38 | 38 | 38 | 38  | 37 = 384 |     |      |        |

The extra day added at the end of Elaphebolion may have been subtracted early in the following month. Or perhaps the months of Mounichion, Thargelion, and Skirophorion had 29, 30, and 29 days, after the reversal in the sequence of alternation with Elaphebolion. Any one of the prytanies from V to VII inclusive may have had the extra day, with the other two at 38 days each. I have shown the sequence above only to illustrate in the conciliar year my belief that there were 167 days from Maimakterion 11 to Elaphebolion 30, a normal number. But the last prytany, or one of the two preceding it, must have had only 37 days. There were slight irregularities in both calendars, perhaps connected with each other, and this could be indicated by writing the sequence of prytanies

39   39   39   39   38   38   39   37   38   38 = 384.

Yet this would be quite uncertain.

332/1 (Ordinary)<br>Archon Niketes

[22] Cf. Pritchett and Neugebauer, *Calendars*, p. 48.

*I.G.*, II², 368 (lines 1-6) [also *I.G.*, II², 344]

a. 332/1 a. ΣΤΟΙΧ. 33

[Προξενία Θε]οφάντ[ου ʽΡοιτειέως· ᾿Επὶ Νική] [23]
[του ἄρχον]τος ἐπὶ τ[ῆς ...ντίδος δευτέρα]
[s πρυτα]νείας ἧι ᾿Αρ[ιστόνους ᾿Αριστόνου<sup>υ</sup>]
[᾿Αναγυ]ράσιος ἐγρα[μμάτευεν· Βοηδρομιῶν]
5 [ος ἐν]άτει ἰσταμέν[ου, δευτέραι καὶ τριακ]
[οστ]ῆι τῆς πρυτανε[ίας· – – – κτλ. – – –]

[Boedromion] 9 = Prytany [II 32] = 68th day.

*Hesperia*, V, 1936, pp. 413-414 (11)

a. 332/1 a. ΣΤΟΙΧ. 31

[θ       ε       ο]       ι
[᾿Επὶ Νικήτου ἄρχοντος ἐπὶ τῆς] Κεκροπί
[δος πέμπτης πρυτανείας ἧι ᾿Αρ]ιστόνου
[s ᾿Αριστόνου ᾿Αναγυράσιος ἐγρ]αμμάτε[υ]
5 [εν· Ποσιδεῶνος ἐνδεκάτει, πέμπ]τει καὶ
[δεκάτει τῆς πρυτανείας· – –] – κτλ. – –

[Posideon 11] = Prytany [V 15] = 158th day.

*I.G.*, II², 345

a. 332/1 a. ΣΤΟΙΧ. 30

θεοί
᾿Επὶ Νικήτου ἄρ[χοντος ἐ]πὶ τῆς ᾿Αντιοχ
ίδος ὀγδόης π[ρυτανεία]s ἧι ᾿Αριστόνο
υς ᾿Αριστόνο[υ ᾿Αναγυράσι]ος ἐγραμ[μά]τ
5 ευεν· ᾿Ελαφη[βολιῶνος ἐν]άτηι ἐπὶ δέ[κ]α,
ἑβδόμ[ηι τῆς πρυτανείας]· – – κτλ. – – –

---

[23] For line 1 see *Polemon*, V, 1952-5, p. 27.

The Fourth Century            87
I.G., II², 346

a. 332/1 a.                ΣΤΟΙΧ. 25

5  ['Επὶ Νικήτου] ἄρχ[ο]ντος ἐπ[ὶ τῆς]
['Αντιοχίδος] ὀγδόης πρυτα[ν]ε[ίας]
[ῆι 'Αριστό]νους 'Αριστόνου 'Αν[αγ]
[υράσιος ἐ]γραμμάτευεν· 'Ελαφ[ηβ]
[ολιῶνος ἐν]άτηι ἐπὶ δέκα, ἑ[β]δόμ
10  [ηι τῆς πρυτα]νείας· – – κτλ. – – –

I.G., II², 347

a. 332/1 a.                ΣΤΟΙΧ. 22

['Επὶ Ν]ικήτου ἄρχοντος [ἐπὶ τ]
[ῆς 'Αν]τιοχίδος ὀγδοίη[ς πρυ]
[τανεί]ας ῆι 'Αριστόνου[ς 'Αρι]
[στόν]ου 'Αναγυράσιος ἐ[γραμ]
5  [μάτε]νεν· 'Ελαφηβολιῶν[ος ἐν]
[άτει] ἐπὶ δέκα, ἑβδόμει τ[ῆς π]
[ρυτα]νείας· – – κτλ. – – – – –

Hesperia, VIII, 1939, pp. 26-27 (6)

a. 332/1 a.                ΣΤΟΙΧ. 29

['Επὶ Νικήτου ἄρχοντος ἐπὶ τῆς 'Αντιο]
[χίδος ὀγδόης πρυτανείας ῆι 'Αριστό]
[νο]υ[ς 'Αριστόνου 'Αναγυράσιος ἔγραμ]
[μά]τευ[ε]ν· 'Ε[λαφηβολιῶνος ἐνάτηι ἐπὶ]
[δέ]κα, ἑβδό[μηι τῆς πρυτανείας· – κτλ. –]

All four of the preceding texts give the equation

Elaphebolion 19 = Prytany VIII 7 = 255th day.

I.G., VII, 4252

a. 332/1 a.                ΣΤΟΙΧ. 27

θεοί

Ἐπὶ Νικήτου ἄρχοντος ἐπὶ τῆς Ἐρε
χθηίδος ἐνάτης πρυτανέας ἧι Ἀρι
στόνους Ἀριστόνου Ἀναγυράσιος
5　ἐγραμμάτευεν· ἐνδεκάτηι, τρίτηι
καὶ εἰκοστῆι τῆς πρυτανείας· - κτλ. –

*I.G.*, VII, 4253

*a.* 332/1 *a.*　　　　　　ΣΤΟΙΧ. 30

θεοί
Ἐπὶ Νικήτου ἄρχοντος ἐπὶ τῆς Ἐρεχθη
ίδος ἐνάτης πρυτανείας ἧι Ἀριστόνο
υς Ἀριστόνου Ἀναγυράσιος ἐγραμμάτ
5　ευεν· Θαργηλιῶνος ἐνδεκάτει, τρίτηι
καὶ εἰκοστῆι τῆς πρυτανείας – κτλ. –

Thargelion 11 = Prytany IX 23 = 306th day.

The year was probably divided into months and prytanies as
follows, and had 354 days. There is no irregularity.

Months　29　30　29　30　29　30　29　30　29　30　29　30 = 354
Prytanies　36　36　36　35　35　35　35　35　35　36 = 354

### 331/0 (Ordinary)
### Archon Aristophanes

*I.G.*, II², 363

*a.* 331/0 *a.*

[Ἐπὶ Ἀριστοφάνους ἄρχοντος]
[ἐπὶ τῆς ....¹⁰..... ἑβδόμ]η[ς πρυταν]　　ΣΤΟΙΧ. 29
[είας ἧι Νικόστρατος Ἀρχέ]λεω [ἐκ Μυρ]
[ρινούττης ἐγραμμάτευε· Ἀ]νθε[στηρι]
5　[ῶνος ἐνδεκάτηι, τρίτηι τῆ]ς πρυ[τανε]
[ίας· ἐκκλησία κυρία· τῶν π]ροέδρω[ν ἐπ]
[εψήφιζεν ......¹³......]εύς vacat

[ἔδοξεν] τῶι δήμω[ι· Πολύευκ]τος Σωσ[τρ]
[άτου Σφ]ήττιος ε[ἶπεν· ἐπει]δὴ Διον[ύσ]
κτλ.

For the continuation of the text, see *I.G.*, II², 363, and *Polemon*, V, 1952-3, p. 27.

Anthesterion 11 = Prytany [VII 3] = 217th day.

I give the text here as a substitute for that suggested earlier by me in *Hesperia*, X, 1941, pp. 47-49, and that more recently proposed by Pritchett and Neugebauer, *Calendars*, p. 55. The difficulties of reconciling this inscription with that of the known text from the archonship of Chremes (326/5) have led them to read the name of the month in lines 4-5 as [Πυα]νο[ψιῶνος], but in fact the upright of what they have taken, or rather restored, as psi is preserved, to the left in its space, as of the letter epsilon, and as given by Lolling in his majuscule copy published in *I.G.*, II, 5, 492 f.

In my earlier study I had suggested that the date should be during the known activity of Polyeuktos (*ca.* 342-318) and further indicated that the year 326/5, where the secretary was unknown, might offer a haven. The reference in the text to a gift of grain from Dionysios suggested help from the tyrant of Herakleia during the famine of 330/326, an interpretation which implied that Dionysios had modified his possible hostility of 330/29 (if this does not read too much into the text of *I.G.*, II², 360, lines 30-45) and become a generous donor in 326/5. If this text is now correctly dated in 331/0, then Dionysios was a donor in that year, though at least inconsiderate of the comfort of ships sailing to Athens in 330/29, which I think need not be held as impossible, or even improbable.

I suggest that the secretary of the year of Aristophanes, known from *I.G.*, II², 348, as Νικόστρατος Λ[- - - - - - - -], should be restored in this text as [Νικόστρατος Ἀρχέ]λεῳ [ἐκ Μυρρινούττης]. The patronymic represents a new reading from the stone. Since I last wrote about the text of *I.G.*, II², 363, I have examined the stone in Athens. The visible letters which I had read (and still do read) from a squeeze as ΑΓΟ should be rather ΛΕΓ⦚. The stone is damaged, and a nick which I had taken as cross-bar of an alpha is, I think, no part of a letter, for it does not extend all the way to the oblique hasta on the right. The second letter seems to me, from the stone, clearly epsilon, and has been so read by others who have seen the stone. I see only the left rounding of the third letter, and read a possible omega. If the letter was omikron, then some patronymic in – – λέο[υ] should probably be supplied.[24] The demotic was long, and for the phyle Aigeis required in 331/0 the restoration ἐκ Μυρρινούττης is unique. The restoration Νικόστρατος Ἀ[ρχέλεω ἐκ Μυρρινούττης] should now be made in *I.G.*, II², 348.[25]

I had also cited for my restoration of the archon's name in line 1 of *I.G.*, II², 363, the parallel of the archon's name in *I.G.*, II², 349, of the year of Aristophanes, with the letters so spaced out as to occupy the whole line.[26] The parallel is all the more striking and appropriate now that we attribute *I.G.*, II², 363, itself to the archonship of Aristophanes.

### *I.G.*, II², 348

a. 331/0 a.

[Ἐπὶ Ἀριστοφάνους]
[ἄρχοντος ἐπὶ τῆς Ἀκα]
μα[ντίδος ὀγδόης, ἐνάτηι]

---

[24] Oikonomides and Koumanoudes, *Polemon*, V, 1952-5, p. 27, read the three letters in this line as ΛΕΥ.

[25] The demotic Κολλυτεύς, tentatively given to Νικόστρατος in *I.G.*, II², iv, p. 8, comes from *I.G.*, II², 350, which Schweigert has assigned to 318/7 (*Hesperia*, VIII, 1939, p. 33). See Pritchett and Neugebauer, *Calendars*, pp. 65-66.

[26] *Hesperia*, X, 1941, p. 49.

ἐπὶ δέ[κα, ἕκτηι τῆς πρυτανείας·]
5 ἐκκλησί[α ἐν Διονύσου· τῶν προέδρων]
ἐπεψήφιζ[εν – – – – – – – – – – – – –]

Νικόστρατος Ἀ[ρχέλεω ἐκ Μυρρινούττης ἐγραμμάτευεν][27]

⟨Elaphebolion⟩ 1[9] = Prytany [VIII 6] = 255th day.

*I.G.*, II², 349

a. 331/0 a.

Ῥηβούλας Σεύθου : ὑὸς Κότυος ἀδελφὸς Ἀνγελ[ῆθεν]
θ        ε        ο        [ί]
Ἐπὶ Ἀριστοφάνους ἄρχοντος
ἐπὶ τῆς Κε[κ]ροπίδος δεκάτης πρυτα        ΣΤΟΙΧ. 28
5 νείας· Σκ[ιρ]οφοριῶνος δεκάτηι ἰσ[τ]
αμένου, [ἕκτ]ει καὶ δεκάτει τῆς πρυ[τ]
ανεία[ς· τῶν] προέδρων – – – κτλ. – –

Skirophorion 10 = Prytany X 1[6] = 335th day.

There is no irregularity of the calendar in this year. The months alternate full and hollow, beginning with hollow Hekatombaion and reversing at Thargelion (which was full), and ending with hollow Skirophorion. The year was ordinary, with 354 days, and had months and prytanies disposed as follows:

| Months | 29 | 30 | 29 | 30 | 29 | 30 | 29 | 30 | 29 | 30 | 30 | 29 = 354 |
|---|---|---|---|---|---|---|---|---|---|---|---|---|
| Prytanies | 36 | 36 | 36 | 36 | 35 | 35 | 35 | 35 | 35 | 35 = 354. | | |

330/29 (Intercalary)
Archon Aristophon

*I.G.*, II², 351 + 624 (cf. *Addenda*)

a. 330/29 a.

[Εὐδήμ]ου Πλαται[έως]

---

[27] See above, p. 90, for the name of the secretary.

[ἐπὶ ᾿Αριστ]οφῶντος ἄρχοντο[ς]　　ΣΤΟΙΧ. 23-25
[ἐ]π[ὶ τῆς] Λεωντίδος ἐνάτη[ς] π[ρυ]
ταν[εία]ς ἧι ᾿Αντίδωρος ᾿Αντί[νου]
5　Παι[ανι]εὺς ἐγραμμάτευεν: ἐ[ν]
[δ]εκά[τ]ηι Θαργηλιῶνος, ἐνάτη[ι]
[κ]αὶ δ[ε]κάτηι τῆς πρυτανείας·
κτλ.

*I.G.*, II², 352

a. 330/29 a.　　　　ΣΤΟΙΧ. 20
[θ]　　ε　　　[ο　　　ί]
[᾿Επὶ ᾿Αριστ]οφῶντ[ος ἄρχον]
[τος ἐπὶ τ]ῆς Λεων[τίδος ἐν]
[άτης πρυ]τανεία[ς ἧι ᾿Αντί]
5　[δωρος ᾿Αν]τίνου Πα[ιανιεὺ]
[ς ἐγραμμ]άτευεν· Θα[ργηλι]
[ῶνος τε]τράδι ἐπὶ δ[έκα, δε]
[υτέραι] καὶ τριακοσ[τῆι τ]
[ῆς πρυτα]νείας· — — κτλ. — —

These two equations must be studied together. The first is reconcilable neither with an ordinary year nor with an intercalary year, and since it conflicts with the second as well, it has been rightly assumed to contain an error. Unger, Mommsen, Kirchner, Dinsmoor, Pritchett and Neugebauer have accepted it as likely that the error lies in the date by prytany, ἐνάτηι καὶ δεκάτηι being taken as a mistake for ἐνάτηι καὶ εἰκοστῆι. It is thus possible to harmonize the two equations and to reconstruct an intercalary year in which the ninth and tenth prytanies had 38 and 39 days and the last two months (whether full and hollow or hollow and full) had 59 days:

Thargelion 11 = Prytany IX ⟨2⟩9 = 337th day
Thargelion 14 = Prytany IX 3[2] = 340th day.

It is questionable whether two meetings of the Ekklesia, with
no indication that either one was in any way exceptional, should
be assumed so close together.[28] The alternative, of course, is to
define the error differently in the first equation, perhaps attribut-
ing it to the date by month rather than to the date by prytany.
However specious the argument that a scribe might have mis-
taken ΔΔΓIIII for ΔΓIIII (so Mommsen, *Chronologie*, p. 466),
and that therefore this was the probable error, it must be said
that error, by its very nature, often defies rational explanation.
Usener thought that the date by month might have been
ἐ[ιρεσιών]ηι Θαργηλιῶνος, as a heortological substitute for νουμηνίαι Θαργηλιῶνος. Whatever plausibility this may once have had
is dissipated by the addition of *I.G.*, II², 624, to the text, showing
that the date as written was actually the eleventh of the month.
But that may have been, none the less, the seat of the error.
If the date by prytany was correct (administratively the more
important), then the date by month may have been the first, i. e.,
Thargelion 1. The equations might then be written:

Thargelion ⟨1⟩ = Prytany IX 19   = 327th day
Thargelion  14 = Prytany IX 3[2] = 340th day.

Mommsen, *Chronologie*, p. 466 note 1, says "An Neumonds-
tagen (νουμηνίαι) hat man nicht gern decretiert." But I know
no formal objection,[29] and prefer such an assumption to that of
closely consecutive meetings of the Ekklesia.

However this may be, it is probable that both equations
should be one day earlier in the year, making an intercalary year
of 384 days:

Thargelion ⟨1⟩ = Prytany IX 19   = 326th day
Thargelion  14 = Prytany IX 3[2] = 339th day.

[28] See above, p. 75.
[29] See H. T. Wade-Gery, *Essays in Greek History* (Oxford, 1958), p. 47
note 2.

In this case three of the first nine prytanies will have had 39 days, the rest 38, and the tenth prytany will have had 39 days. I regard such an arrangement as entirely normal and preferable to assuming a conciliar year of 385 days. Months and prytanies may be disposed as follows:

Months     30 29 30 29 30 29 30 29 30 29 30 29 30 = 384
Prytanies  39 39 39 38 38 38 38 38 38 39 = 384.

### 329/8 (Ordinary)
### Archon Kephisophon

#### *I.G.*, VII, 4254

a. 329/8 a.                    ΣΤΟΙΧ. 27

θ        ε        ο        ι
Ἐπὶ Κηφισοφῶντος ἄρχοντος ἐπὶ τ
ῆς Ἱπποθωντίδος τρίτης πρυτανε
ίας ἧι Σωστρατίδης Ἐχφάντου Εὐπ
5  ρίδης ἐγραμμάτευεν·  *υυυυυυυυυ*
ἕκτει ἐπὶ δέκα, τρίτει καὶ τριακο
στεῖ τῆς πρυτανείας· ἐκκλησία· τῶ
κτλ.

⟨Pyanepsion⟩ 16 = Prytany III 33 = 105th day.

Space was left on the stone (though not enough) for the name of the month.

#### *I.G.*, II², 353

a. 329/8 a.                    ΣΤΟΙΧ. 23

[Ἐπὶ] Κηφισο[φῶντος ἄρχοντος]
[ἐπ]ὶ τῆς Αἰγε[ῖδος τετάρτης π]
[ρ]υτανείας ἧι [Σωστρατίδης Ἐ]
[χ]φ[ά]ντου Εὐπυ[ρίδης ἐγραμμά]
5  [τευ]εν· Πυανοψ[ιῶνος ἕνηι καὶ]

$$[\nu\epsilon\alpha]\iota, \; \epsilon\nu\delta\epsilon\kappa\alpha[\tau\eta\iota \; \tau\eta s \; \pi\rho\upsilon\tau\alpha\nu\epsilon]$$
$$[\iota]\alpha[s\cdot] \; \tau\omega\nu -- [ \; -- \kappa\tau\lambda. \; --]$$

Either

Pyanepsion [30] = Prytany [IV] 11 = 119th day

or

Pyanepsion [29] = Prytany [IV] 11 = 118th day.

The first equation is correct if Pyanepsion was full and the third prytany had 36 days; the second is correct if Pyanepsion was hollow and the third prytany had 35 days. With their strict interpretation of Aristotle's rule of prytanies Pritchett and Neugebauer naturally hold to the first equation; I prefer the second, in order to avoid too many months of 30 days each in close succession and in order to hold the count of days in the ordinary year down to 354 (astronomically the better solution).[30]

There is evidence in *I.G.*, II², 1672, that in this year the first and second prytanies were of 36 days each and the fifth and sixth of 35 days each, and that Hekatombaion was full.[31] Months and prytanies within the year may be arranged as follows:

| Months | 30 | 29 | 30 | 29 | 30 | 29 | 30 | 29 | 30 | 29 | 30 | 29 = 354 |
|---|---|---|---|---|---|---|---|---|---|---|---|---|
| Prytanies | 36 | 36 | 35 | 35 | 35 | 35 | 35 | 35 | 36 | 36 = 354. | | |

328/7 (Intercalary) [32]

Archon Euthykritos

*I.G.*, II², 452 (cf. *A.J.P.*, LIX, 1938, p. 499;
Pritchett and Meritt, *Chronology*, p. 2) [33]

---

[30] Parker and Dubberstein, *Babylonian Chronology*, p. 36, show visibility of the new crescent in 329 B.C. on July 18 and in 328 B.C. on July 7.

[31] See Pritchett and Neugebauer, *Calendars*, pp. 36, 51 with note 30.

[32] This year was taken as ordinary by Pritchett and Neugebauer, *Calendars*, pp. 51-52.

[33] The stone permits the reading [Χολ]λείδης in line 8. In line 10 the name and demotic of the last symproedros were proposed by D. M. Lewis, *B.S.A.*, XLIX, 1954, p. 49 (cf. *S.E.G.*, XIV, 56); there must have been two uninscribed spaces at the end of the line. In line 11 the first 16 letters in the name of

a. 328/7 a.                                ΣΤΟΙΧ. 36

['Επὶ Εὐθυκρί]του ἄρχοντος
[ἐπὶ τῆς 'Ακαμαντίδ]ος ἕκτης πρυτανε[ίας ᵛ ἧι Π]
[υθόδηλος Πυθοδήλ]ου 'Αγνούσιος [ἐ]γρ[αμμάτευ]
[ε· Γαμηλιῶνος ὀγδό]ει ἐπὶ δέκα, μιᾶι κ[αὶ τριακ]
5   [οστεῖ τῆς πρυτανε]ίας· ἐκκλη[σία κυρία· τῶν πρ]
[οέδρων ἐπεψήφιζε]ν Πάμφιλο[ς] Φ[ηγούσιος· συμ]
[πρόεδροι ....⁸.... 'Α]λαιεύς, Ο[ – – Pandionidis –]
[– –, – – – – – – Χολ]λείδης, Α[ – – Oeneidis – –]
[– – –, –¹⁵– – – – Μ]ελιτεύς, Π[– –¹² – – Hippo]
10  [thontidis, –¹³– Οἰν]αῖος, Βουλ[ις Θοραιεύς ᵛᵛ]
[Λυκοῦργος Λυκόφρο]νος Βουτά[δης εἶπεν· ...]
[................¹⁷.........] ⌒ΣΕΓΙΜΣ[............]
[...................]ΣΓ–Λ[.............]

[Gamelion] 1[8] = Prytany VI [3]1 = 225th day.

The date within the prytany is certainly the 31st. Koehler's
readings were correct, and should not have been challenged by
Premerstein and Kirchner. I examined the stone in 1958 and
found every letter of μιᾶι sure, and the vertical stroke of the
kappa in κ[αί] which follows.³⁴

*I.G.*, II², 354

a. 328/7 a.                                ΣΤΟΙΧ. 34

[θ          ε]          ο          ί
['Επ' Εὐθυκρίτου ἄρ]χοντος, ἱερείως δὲ 'Ανδρο
[κλέους ἐκ Κεραμ]έων· ἐπὶ τῆς 'Αντιοχίδος ὀγ
[δόης πρυτανεία]ς ἧι Πυθόδηλος Πυθοδήλου

Lykourgos occupied seventeen letter-spaces. The text here given differs
from those earlier proposed.
³⁴ One may disregard the discussion of this line in the *Corpus* and in Pritchett
and Neugebauer, *Calendars*, p. 51. Klaffenbach, in *Gnomon*, XXI, 1949, p. 135,
writes: "Den Angaben der Verf. bezüglich des Abklatschbefundes von IG
II/III² 452 stimme ich nach eigener Prüfung vollkommen bei; nur δέκα ist
gesichert." Another testimony to the occasional unreliability of a squeeze!

5  [Ἀγνούσιος ἐγρα]μμάτευεν· ἔνηι καὶ νέαι π
   [ροτέραι ⟨πέμπτει καὶ⟩ εἰκοστ]ῆι τῆς πρυτανείας· ἐκκλησ
   [ία· – – – – – κτλ. – – – – –] – – – – – –

⟨Elaphebolion⟩ 29 = Prytany VIII [2⟨5⟩] = 295th day.

The letter at the end of line 5, which is critical in determining
the nature of the calendar, is pi, not epsilon. From examination
of both squeeze and stone, I concur fully in the findings of
Pritchett and Neugebauer and with the reading as given in the
*Corpus*. In an intercalary year the last day of Elaphebolion,
if planned as a hollow month, falls on the 295th. This should
be equated with the 25th day of the eighth prytany. And,
indeed, one might restore π[έμπτηι καὶ εἰκοστ]ῆι, assuming that
the scribe twice violated stoichedon order by inscribing two
letters in the space of one. Pritchett and Neugebauer assume
one such violation and write π[έμπτει καὶ δεκάτ]ηι. But this
cannot be reconciled with the equation of *I.G.*, II², 452, and is
no longer tenable. There is, I think, a better way: to assume
that the date by month was ἔνηι καὶ νέαι π[ροτέραι] and that
the scribe simply omitted πέμπτηι καὶ when he meant to write
πέμπτηι καὶ εἰκοστῆι. One may assume that in his imagination
he had finished writing πέμπτηι καί when he was only at the
end of προτέραι, being confused by the similarity in the endings
of the words. Elaphebolion, originally planned for 29 days,
was given an extra day and made 30, reversing the sequence of
alternation. The months and prytanies in the year were disposed as follows:

| Months | 30 | 29 | 30 | 29 | 30 | 29 | 30 | 29 | 30 | 30 | 29 | 30 | 29 | = 384 |
|---|---|---|---|---|---|---|---|---|---|---|---|---|---|---|
| Prytanies | 39 | 39 | 39 | 39 | 38 | 38 | 38 | 38 | 38 | 38 | | | | = 384. |

Except for an oversight on the part of the stonecutter of
*I.G.*, II², 354, there is no irregularity.

327/6 (Ordinary) [35]

Archon Hegemon

*I.G.*, II², 357

*a.* 327/6 *a.*                                    ΣΤΟΙΧ. 29

Προξενί[α — — — — — — — — — — — — 'Ερε]
τριεῖ· 'Επ[ὶ 'Ηγήμονος ἄρχοντος ἐπὶ τῆ]
ϛ Αἰαντ[ίδος ἕκτης πρυτανείας ἧι Αὐ]
τοκλῆς [Αὐτίου 'Αχαρνεὺς ἐγραμμάτε]
5  νεν· ἔνηι κ[αὶ νέαι, τρίτηι τῆς πρυταν]
εἰας· — — — — — κτλ. — — — — — —

⟨Posideon⟩ 30 = Prytany [VI 3] = 178 day.

The stoichedon order requires that the prytany be the sixth,
and in this prytany the first day can be equated with the last
day of a month (in this case Posideon, though the name was
omitted) if the months and prytanies were ordered as follows:

30   29   30   29   30    30 = 178 days
36   36   35   35   35 +   1 = 178 days.

But if the prytanies each had 35 days, then the restoration of
the date by prytany can be made as [τρίτηι τῆς πρυταν]εἰας,
which I prefer because of the better correlation with *I.G.*, II²,
356, of the following prytany.

*I.G.*, II², 356

*a.* 327/6 *a.*                                    ΣΤΟΙΧ. 20

['Επὶ 'Ηγήμ]ονος ἄρχοντο[ς ἑ]
[πὶ τῆς 'Ιπ]ποθων[τί]δ[ος ἑ]β[δ]
[όμης πρυ]τανεία[ς ἧι Αὐτο]

---

[35] Pritchett and Neugebauer, *Calendars*, pp. 52-54, treated this year as intercalary. Pritchett, *B.C.H.*, LXXXI, 1957, p. 282 note 1, has expressed his
intention of returning to the calendar problems of this year.

$[\kappa\lambda\hat{\eta}s \; A]\dot{v}\tau\acute{\iota}ov \; [\text{'}A\chi]\alpha\rho[\nu]\epsilon[\dot{\upsilon}s] \; \dot{\epsilon}[\gamma]$
5  $[\rho\alpha\mu\mu]\acute{\alpha}\tau\epsilon\nu\epsilon\nu\cdot \; \delta\epsilon\upsilon\tau[\acute{\epsilon}\rho\alpha\iota] \; \phi\theta$
   $[\acute{\iota}\nu]o\nu\tau os, \; \ddot{\epsilon}\kappa\tau\epsilon\iota \; [\kappa]\alpha[\grave{\iota} \; \epsilon]\ddot{\iota}[\kappa]o\sigma$
   $[\tau\hat{\eta}]\iota \; \tau\hat{\eta}[s \; \pi]\rho\upsilon\tau\alpha\nu\epsilon\acute{\iota}\alpha[s \; - \; - \; -]$

⟨Anthesterion⟩ 29 = Prytany [VII] 26 = 236th day.[36]

The months and prytanies can be arranged as follows:

| Months | 30 | 29 | 30 | 29 | 30 | 30 | 29 | 30 | 29 | 30 | 29 | 30 = 355 |
|---|---|---|---|---|---|---|---|---|---|---|---|---|
| Prytanies | 35 | 35 | 35 | 35 | 35 | 35 | 36 | 36 | 36 | 37 = 355. | | |

In case the Athenians found at the end of the year that Hekatombaion 1 was due to fall too late after the observable crescent
of the moon, then the shortening of the year by one day would
have left an ordinary year of 354 days, in which the prytanies
were arranged in order in a sequence exactly the opposite of
that specified by Aristotle. I hold that this, even so, does not
violate the spirit of Aristotle's law, and I hold the same to be
true if the year had 355 days and one of the last prytanies then
happened to have 37 days.[37] The alternatives to which one has
resorted to save the letter of the law seem to me much more
difficult: the assumption of a continuous retardation of the festival calendar by approximately three days in the middle of the
year, with the corollary that the Athenians (who by hypothesis
regulated their calendar by the observed crescent) disregarded
the crescent at the ends of at least three consecutive months.
Surely, whatever advantage anyone gained by a posited temporary retardation need not have prolonged the maladjustment in
this fashion after the crisis was past. A grave disadvantage in

---

[36] For lines 2-3 of the text see Dinsmoor, *Archons*, p. 371; for lines 3-4 see
Meritt, *Hesperia*, III, 1934, p. 4.
[37] Compare the sequences in the third and second centuries, where sometimes
six prytanies of 30 days were followed by six prytanies of 29 days, and where,
conversely, sometimes six prytanies of 29 days were followed by six prytanies
of 30 days. See below, pp. 130, 134-144.

the schemes for maintaining a rigid Aristotelian rule is the frequent necessity for assuming long periods of maladjustment in the festival calendar. It is paradoxical to start from a hypothesis of continuous observation of the lunar crescent for each new moon and then to allow such frequent and complete disregard of the assumed observations.

There is another text from the archonship of Hegemon, which probably should not be used in reconstructing the calendar, except to show that Μουνιχιῶνος ἐνάτη μετ᾽ εἰκάδας fell in the ninth prytany. I offer here, however, a tentative new restoration:

Hesperia, III, 1934, pp. 3-4 (5)

a. 327/6 a.                  ΣΤΟΙΧ. 33

[’Εφ᾽ Ἡγήμονος ἄρχοντ]ος ἐπὶ τῆς Οἰνηί[δος ἐ]
[νάτης πρυτανείας] ἧι Αὐτοκλῆς Αὐτί[ου ’Αχ]
[αρνεὺς ἐγραμμάτε]νεν· Μουνιχιῶνος· [πέμπ]
[τει τῆς πρυτανεία]ς̣· ἐνάτει μετ᾽ εἰκά[δας· τ]
5 [ῶν προέδρων ἐπεψή]φιζεν Πάμφιλος Π̣[αιαν]
[ιεύς· ἔδοξεν τῶι δή]μωι    vacat

Mounichion 22 = Prytany [IX 5] = 288th day.

The equation is possible with months and prytanies arranged as follows:

| Months | 30 | 29 | 30 | 29 | 30 | 30 | 29 | 30 | 29 | 30 | 29 | 30 = 355 |
|---|---|---|---|---|---|---|---|---|---|---|---|---|
| Prytanies | 35 | 35 | 35 | 35 | 35 | 35 | 36 | 37 | 36 | 36 = 355. | | |

I have recently examined the stone in Athens. Pritchett and Neugebauer are entirely justified in their assertion that the letter before ἐνάτει in line 4 cannot be iota.[38] My first suggestion[39] that the lacuna in lines 3-4 beteen Μουνιχιῶνος and ἐνάτει μετ᾽ εἰκά[δας] should be filled with the restoration ἐκκλησία ἐν

[38] Calendars, p. 53.
[39] Hesperia, III, 1934, p. 3.

τῶι θεάτρωι is not sound. Nor do I believe that the alternative now offered is desirable: Μουνιχιῶνος [δευτέραι, ἡμερολεγδὸν] δ᾽ ἐνάτει μετ᾽ εἰκά[δας]. It is claimed that the letter preceding ἐνάτει is almost certainly a delta, and delta is read with no dot underneath. But what is preserved is, in fact, only the right-hand tip of a low horizontal stroke, which in this inscription could belong to sigma, epsilon, or perhaps delta (of which no example is preserved).

Normally there are four elements in the dating of a decree: (1) name and number of the prytanizing phyle, (2) name of the month, (3) date within the month, and (4) date within the prytany. In the other two decrees of the archonship of Hegemon the scribe omitted the name of the month. My belief is that here he added the name of the month but changed the normal order. When such dates occur in the fifth century the normal order is (1) (2) (3) (4) in *I.G.*, I², 324, for example,[40] or (1) (4) (3) (2) in *I.G.*, I², 304B.[41] What I posit for the present text is the order (1) (2) (4) (3).[42] The year being ordinary, the 22nd day of Mounichion (ἐνάτη μετ᾽ εἰκάδας) may be equated with the fifth day of the ninth prytany, and I have so restored it.[43]

### 326/5 (Ordinary)
### Archon Chremes

For the text of *I.G.*, II², 359, see above, p. 7. There is no

---

[40] See Meritt, *Athenian Financial Documents*, p. 140 (lines 57-58) and p. 141 (lines 78-79).

[41] See Meritt, *Athenian Financial Documents*, pp. 119-122 *passim*.

[42] One may note that this is the order in the prescripts of the text published in *Hesperia*, XVII, 1948, pp. 25-26 (12, lines 1-4 and 38-41), if one omits only the date κατὰ θεόν within the month.

[43] Arnold Gomme, *Cl. Rev.*, LXIII, 1949, p. 122, has also found the restoration suggested by Pritchett and Neugebauer difficult: "I cannot understand the restoration δευτέραι (in the inscription, *Hesp.* iii (1934), p. 3) to mean 22nd of the month, on p. 53. This last carries with it the further restoration, ἡμερολεγδόν, for which the authors suggest a new and not very convincing interpretation."

irregularity. The calendar equation is:

Elaphebolian 8 = Prytany [VII] 30 = 244th day.

Months    29  30  29  30  29  30  29  30  29  30  29  30 = 354
Prytanies  36  36   36  36  35  35  35  35  35  35 = 354.

325/4 (Intercalary)
Archon Antikles

*I.G.*, II², 360

*a.* 325/4 *a.*                                    ΣΤΟΙΧ. 39 (lines 1-2)

Ἐπ᾽ Ἀντικλέους ἄρχοντος ἐπὶ τῆς Αἰγεῖδος πέμπτ
ης πρυτανείας ἧι Ἀντιφῶν Κοροίβου Ἐλευσί ἔγρα
μμάτευεν· ἑνδεκάτηι, τετάρτηι καὶ τριακοστῆι τῆς πρυταν
είας – – – – – – – – – – – κτλ. – – – – – – – – –

⟨Posideon II⟩ 11 = Prytany V 34 = 188th day.

*I.G.*, II², 361

*a.* 325/4 *a.*                                    NON-ΣΤΟΙΧ.

[Ἐπ᾽ Ἀντι]κλείους ἄρχοντος ἐπὶ τῆς Ἀκαμαν[τίδ]
[ος δ]εκάτης πρυτανείας ἧι Ἀντιφ[ῶν Κο]
[ρ]οίβου Ἐλευσίνιος ἐγραμμάτευεν· Θ[αργηλιῶ]
νος ὀγδόηι μετ᾽ εἰκάδας, πέμπτ[ηι τῆς πρυτα]
5   [ν]είας· – – – – – – – κτλ. – – – – – – – – –

Thargelion 22 ⟨κατ᾽ ἄρχοντα, 25 κατὰ θεόν⟩ =
Prytany X 5 = 350th day.

Here Pritchett and Neugebauer (*op. cit.*, pp. 55-56), with
a rigid prytany calendar, find either that they must assume, of
the first six months, that five were full and one hollow (an
anomaly which they rightly reject) or that "the archons intercalated two or three extra days in the first part of the year.
– – – – this intercalation will explain, in turn, the shortening of

the year at the end of Skirophorion, as attested by the second equation, which indicates that the year had on Thargelion 23 [*sic*] only 33 or 34 days to run."

I have nowhere found any evidence that extra days were ever dropped from Skirophorion. In this particular instance, if the equation Thargelion 22 = Prytany X 5 were really correct, I should argue for a final prytany of 42 (or 41) days.[44] But the study made by Pritchett and Neugebauer of dates κατ᾽ ἄρχοντα and κατὰ θεόν allows an assumption of less drastic irregularity. The dates as given by month are sometimes not so late as in reality they ought to be. This is the significance of dates κατ᾽ ἄρχοντα (dates prior to that on the stone have been added, and the calendar count is retarded) as opposed to dates κατὰ θεόν. When the secretary thought it worth while he sometimes indicated the correct date, as well as the tampered date. But he did not necessarily do this, and he is not known to have done it ever so early as the fourth century. It appears from our study so far that the calendar counts as we have them are by and large correct. Irregularities were apparently introduced, if at all, for some special whim or purpose, and—the purpose served— the irregularity was erased without long delay. That two or three days intercalated into the year 325/4 before Posideon II 11 should have been allowed to vitiate the approximate coincidence of each succeeding first of the month with the phenomenon of the new lunar crescent, an observable and observed error perpetuated throughout the remaining six months of the year, and serving no useful purpose once the temporary occasion for the interposition was past, is more intolerable than deviation from a rule of prytanies which needs to be taken as rigid only because a hypothesis of modern scholarship seeks to make it so.[45]

I hold, therefore, that there is no irregularity in the calendar

[44] See *Hesperia*, IV, 1935, p. 536.                    [45] See below, p. 116.

during the first six months. If the first two prytanies had 39 days each and the next two 38 days each, then a regular alternation of full and hollow months brings Posideon II 11 to Prytany V 34 (the 188th day).[46] Nor do I assume any irregularity in the calendar before Thargelion, though our latest investigators have here assumed perpetual irregularity. The addition of extra days late in Thargelion may have been occasioned by a desire to postpone the celebration of the Plynteria,[47] or for whatever reason. We can now hardly more than guess. But when the occasion passed, one expects the calendar to revert to normal, and I propose that this was managed before the end of the month. Thargelion, therefore, had 29 days, 25 of them falling before and including Thargelion 22 and the span from Thargelion 23 to 29 inclusive having only four of them.[48]

I arrange the months and prytanies of the year 325/4 as follows:

Months    30 29 30 29 30 29 30 29 30 29 30 29 30 = 384
Prytanies 39 39 38 38 38 38 38 38 39 39 = 384.

### 324/3 (Ordinary)
### Archon Hegesias

*Hesperia*, X, 1941, pp. 49-50 (12)

As published in *Hesperia* this text shows the unusual form ἐγραμμάτευε restored in line 4 without the final nu. Though possible,[49] this is less likely, and the better restoration is with

---

[46] The sequence of prytanies is like that in 336/5 B.C. See above, p. 15.

[47] See L. Deubner, *Attische Feste*, pp. 17-18.

[48] I take it that the dates μετ' εἰκάδας, in a manner analogous to the dates φθίνοντος when δεκάτη ὑστέρα meant 21st, omitted in a hollow month the date ἐνάτη μετ' εἰκάδας when the count was backward. When the count was forward the date omitted was likewise ἐνάτη μετ' εἰκάδας (see above, pp. 58-59). Forward count is relatively rare in the inscriptions (see above, p. 51). The change from a normal forward count to a normal backward count must have taken place before the epigraphical evidence is available.

[49] See, for example, the text immediately following.

ἐγραμμάτευεν. Probable names of the month are now reduced to two: Posideon and Gamelion. This indicates the date as in the fifth prytany, and hence with the phyle Aiantis (*op. cit.*, p. 50). The text may be restored:

a. 324/3 a. ΣΤΟΙΧ. 25

['Επὶ Ἡγησίου ἄρχοντος ἐπ]ὶ τῆς Α
[ἰαντίδος πέμπτης πρυτα]νείας
[ἧι Εὐφάνης Φρύνωνος Ῥαμ]νούσι
[ος ἐγραμμάτευεν· Ποσιδε]ῶνος<sup>v</sup>

  5  [τρίτηι μετ' εἰκάδας, μιᾶι κ]αὶ τρ
[ιακοστῆι τῆς πρυτανείας· ἐ]κκλ
[ησία· – – – – – – – – – – – – –]

[Posideon 28] = Prytany [V] 3[1] = 175th day.

The same calendar equation would be obtained if the date by month was ὀγδόηι μετ' εἰκάδας with forward count.

*I.G.*, II², 362 (see *Hesperia*, X, 1941, p. 47)

a. 324/3 a. ΣΤΟΙΧ. 28

['Ε]φ' Ἡγησίου ἄρχ[οντος ἐπὶ τῆς 'Ακαμα]
[ν]τίδος ἐνάτης [πρυτανείας ἧι Εὐφά]
[νη]ς Φρύνωνος Ῥ[αμνούσιος ἐγραμμά]
[τ]ενε· Θαργηλ[ιῶνος ὀγδόηι ἐπὶ δέκα,]

  5  [ἐ]νάτει καὶ ε[ἰκοστεῖ τῆς πρυτανεία]
[ς·] ἐκκλησί[α κυρία· τῶν προέδρων ἐπε]
[ψ]ήφιζεν – – – – – – – – – – – – – –

Thargelion [18] = Prytany IX 29 = 313th day.

*I.G.*, II², 547 (cf. Pritchett and Meritt, *Chronology*, p. 2)

['Εφ' Ἡγησίου ἄρχοντος ἐπὶ τῆς 'Ερεχθ]
[εῖδος δεκάτης πρυτανεία]ς ἧ[ι Εὐφά]
[νης Φρύνωνος Ῥαμνούσι]ος ἐγ[ραμμά]

[τευεν· Θαργηλιῶνος] δευτέρα[ι μετ᾽ ε]
5  [ἰκάδας, πέμπτηι τῆς] πρυτανε[ίας· ..]
κτλ.

[Thargelion] [2]9 = Prytany [X 5] = 324th day.

*I.G.*, II², 454 (see *A.J.P.*, LIX, 1938, p. 499)

*a.* 324/3 *a.*                                  ΣΤΟΙΧ. 38

['Εφ᾽ Ἡγησίου ἄρχοντος ἐπὶ τῆς 'Ερε]χθηίδος δεκάτ
[ης πρυτανείας ἧι Εὐφάνης Φρύνων]ος 'Ραμνούσιο
[ς ἐγραμμάτευεν· Σκιροφοριῶνος ἔκ]τει μετ᾽ εἰκά
[δας· ἐκκλησία· – – – – – – – – – –] – – κτλ. – –

This yields no calendar equation.

The equations in this year present no irregularities, and months and prytanies may be arranged as follows:

Months      29  30  29  30  29  30  30  29  30  29  30  29 = 354
Prytanies   36  36  36  36  35  35  35  35  35  35 = 354.

### 323/2 (Ordinary)
### Archon Kephisodoros

*I.G.*, II², 365

*a.* 323/2 *a.*                                  ΣΤΟΙΧ. 32

'Επὶ Κηφισοδώρου ἄρχον[τος ἐπὶ τῆς 'Ιππο]
θωντίδος πρώτης πρυτα[νείας ἧι 'Αρχίας]
[Π]υθοδώρου ἐγραμμάτευε[ν· 'Εκατομβαιῶν]
[ο]ς· ἑνδεκάτει τῆς πρυτανεί[ας· – – –]

[Hekatombaion] ⟨11⟩ = Prytany I 11 = 11th day.

*I.G.*, II², 366

*a.* 323/2 *a.*  NON-ΣΤΟΙΧ.

'Επὶ Κηφισοδώρου

ἄρχοντος ἐπὶ τῆς Αἰγ
ηίδος δευτέρας πρυτα
νείας· ἕκτηι καὶ δεκά
5 τηι τῆς πρυτανείας
κτλ.

There is no equation.

### I.G., II², 367

a. 323/2 a.      ΣΤΟΙΧ. 26

['Επὶ Κηφισοδώρου] ἄρχοντος ἐπὶ ᵛ
[τῆς ...⁶... ίδος τ]ρίτης πρυτανε
[ίας ἧι 'Αρχίας Πυθ]οδώρου 'Αλωπεκ
[ῆθεν ἐγραμμάτευ]εν· Πυανοψιῶν[ο]
5 [ς ἐνάτηι ἐπὶ δέκα,] ἕκτει καὶ τρια
[κοστεῖ τῆς πρυτα]νείας· – – κτλ. –

Pyanepsion [19] = Prytany III 36 = 108th day.

It will appear below, from the equation in *I.G.*, II², 448, that
the year was surely ordinary. I return, therefore, in principle,
to the restoration of the calendar as given in the *Corpus*. The
suggestion offered by Pritchett and Neugebauer,[50] Πυανοψιῶν[ος
ἕκτει φθίνοντος], is scaled to an intercalary year, and is also too
long by one letter for the stoichedon spacing.

### I.G., II², 368 (lines 19-24)

a. 323/2 a.      ΣΤΟΙΧ. 33

['Επὶ Κηφι]σοδώ[ρου ἄρχοντος ἐπὶ τῆς Πανδι]
20 [ονίδος πέ]μπ[της πρυτανείας ἧι 'Αρχίας Πυ]
[θοδώρου] 'Αλω[πεκῆθεν ἐγραμμάτευεν· Ποσι]
[δεῶνος] δευτ[έραι ἱσταμένου ἐμβολίμωι, ὁ]
[γδόει τῆ]ς π[ρυτανείας· τῶν προέδρων ἐπεψ]

108                    The Fourth Century

[ἤφιζεν . ]ω[– – – – – – – – – – – – – – – – –]

[Posideon] 2[+1] = Prytany V [8] = 151st day.

*I.G.*, II², 448 (lines 1-5)

a. 323/2 a.                                   ΣΤΟΙΧ. 41

Ἐπὶ Κηφισοδώρου ἄρχοντος ἐπὶ [τῆς Πανδιονίδος πέ]
νπτης πρυτανείας ἧι ['Αρ]χί[ας] Πυ[θοδώρου Ἀλωπεκῆθε]
ν ἐγραμμάτευεν· Ποσιδεῶνος ἕκ[τηι ἐπὶ δέκα, δευτέρ]
αι καὶ εἰκοστεῖ τῆς πρυτανεία[ς· ἐκκλησία κυρία· τῶ]
5  ν προέδρων – – – – – – κτλ. – – [– – – – – – – –]

Posideon [1]6 = Prytany V 2[2] = 165th day.[51]

The equations of *I.G.*, II², 368 and 448, as given above, must
be studied together. Since reading the comments made by
Pritchett and Neugebauer on the text of *I.G.*, II², 448,[52] I have
examined the stone in Athens. The word that follows Ποσιδεῶνος in line 3 is ἕκ[τηι],[53] and the equation determines an ordinary
year.

The difficulty with the calendar in this year has been not with
*I.G.*, II², 448, but with *I.G.*, II², 368. Pritchett and Neugebauer
noted that the usual restorations made room for ἐκκλησία κυρία
in both texts, and they have believed that these "restorations
for the assemblies are the only ones which fit the stoichedon
order."[54]

I suggest that all reference to the Ekklesia be omitted from

[51] E. Schweigert has shown that *I.G.*, II², 343, was probably passed on the
same day (*Hesperia*, IX, 1940, p. 343).
[52] *Calendars*, pp. 58-59.
[53] I am indebted to Ch. Karouzos, Director of the National Archaeological
Museum, for opening a closed section of the museum and permitting me to
examine this stone (Γλυπτά 1482). The kappa in the word ἕκ[τηι], about which
there has been doubt, is entirely clear—all three strokes—and was seen and
confirmed by Mastrokostas, Papademetriou, and Mitsos, as well as by me.
I was not able to make out any sure trace of a following letter.
[54] *Calendars*, p. 58.

*I.G.*, II², 368, as was done, for example, in *I.G.*, II², 380 and 388, just a few years later. This avoids the anomaly of two ἐκκλησίαι κυρίαι in the same prytany, and solves one of the epigraphical problems without recourse to violation of stoichedon order.

Since the equation of *I.G.*, II², 448, shows that Prytany V 22 fell on Posideon 16 one can count back (within the same month and same prytany) to find that Prytany V 9 should have fallen on Posideon 3, and possibly Prytany V 8 on an intercalated Posideon 2. This equation can be written into the text of *I.G.*, II², 368, as above, with no violation of stoichedon order.

*Hesperia*, IX, 1940, p. 336

There is no equation.

The equations in the month Posideon now show that an extra day was intercalated after Posideon 2. This probably does not introduce an irregularity into the calendar, for there was no compensation before Posideon 16, which was the 165th (rather than the 164th) day. The month Posideon was planned originally as a month of 29 days, but the archon must have decided that the months were beginning unduly far ahead of the observed crescent, and this abnormality he rectified by intercalating a day.[55]

The months and prytanies should probably be arranged as follows:

| Months | 30 | 29 | 30 | 29 | 30 | 30 | 29 | 30 | 29 | 30 | 29 | 30 = 355 |
|--------|----|----|----|----|----|----|----|----|----|----|----|-----------|
| Prytanies | 36 | 36 | 36 | 35 | 35 | 35 | 35 | 35 | 36 | 36 = 355. | | |

One notes that the days of Posideon, though 30 in number,

---

[55] The lunar months actually averaged longer than usual during the first half of the year 323/2. Parker and Dubberstein, *Babylonian Chronology*, p. 36, show four lunations of 30 days and only two of 29 days. It is quite possible that we have here an example of that occasional *ad hoc* correction in the system of alternating months which I believe to be characteristic of the Athenian festival calendar.

110                    The Fourth Century

will all show after Posideon 2 a denomination at variance by one
from actuality. If one wishes, he may refer to Posideon 16 in
*I.G.*, II², 448, for example, as Posideon 17 κατὰ θεόν. And so on
to the end: the month ended as a full month κατὰ θεόν.

322/1 (Intercalary)
Archon Philokles

*I.G.*, II², 376
a. 322/1 a.                                    ΣΤΟΙΧ. 31

[ Ἐπὶ Φιλοκλέο]υς ἄρχοντο[ς ἐπὶ τῆς ....]
[.....¹¹.....η]ς πρυτανε[ίας ἧι Εὐθυγέ]
[νης Ἡφαιστοδήμ]ου Κηφι[σιεὺς ἐγραμμά]
[τευεν· ...⁶... ὦνο]ς ἔνη[ι καὶ νέαι...⁵..]

The month was either Posideon or Gamelion, i. e., in an
intercalary year either the sixth or the eighth month. The last
day of the eighth month would fall in the seventh prytany, the
last day of the sixth month in the fifth. There is no calendar
equation and no evidence here for the character of the year.

*I.G.*, II², 372 (EM 7184; cf. Schweigert,
*Hesperia*, VIII, 1939, p. 174)

a. 322/1 a.                                    ΣΤΟΙΧ. 27

[ Ἐπὶ Φιλοκλέους ἄρχοντος ἐπὶ τῆς]
[...ντίδος ὀγ]δό[ης πρυτανείας ἧι]
[Εὐθυγένης Ἡφ]αισ[τοδήμου Κηφισι]
[εὐς ἐγραμμάτ]ευε[ν· Ἐλαφηβολιῶνο]
5  [ς τρίτει ἐπὶ] δέκα, [ὀγδόηι τῆς πρυτ]
[ανέας· ἐκκλ]ησία [ἐν Διονύσου· τῶν π]
[ροέδρων ἐ]πεψήφ[ιζεν .....¹⁰.....]
[.....⁷...]ς· ἔδοξ[εν τῶι δήμωι ᵛᵛᵛᵛᵛ]
[Δημάδη]ς Δημέ[ου Παιανιεὺς εἶπεν·]
10 [ἐπειδὴ Λ]υκο[- - - - - - - - - -]

[Elaphebolion] 1[3] = Prytany VIII [8] = 278th day.[56]

The reconstruction of the year as intercalary is indicated by the fact that the two preceding years were ordinary and the next but one following (320/19) intercalary.

However the next two equations in the year of Philokles are taken, it is obvious that they both involve error on the stone. This has long been recognized. The error in *I.G.*, II², 373, is difficult to define, but that in *I.G.*, II², 375, is clear-cut, and easily corrected: the month was written Thargelion when it should have been Skirophorion.

I.G., II², 373 (lines 16-19)

*a.* 322/1 *a.*                                      ΣΤΟΙΧ. 39

['Επ]ὶ Φιλοκλέους ἄρχοντος ἐπὶ τῆς Οἰνεῖδος ἐνά[τ]
[ης] πρυτανέας ἧι Εὐθυγένης Ἡφαιστοδήμου Κηφι[σ]
[ιε]ὺς ἐγραμμάτευεν· Θαργηλιῶνος δευτέραι ἱστα
[μέ]νου, τρίτει καὶ εἰκοστεῖ τῆς πρυτανείας· – κτλ.

I.G., II², 375

*a.* 322/1 *a.*                                      ΣΤΟΙΧ. 25

['Επὶ Φι]λοκ[λέους ἄρχοντος ἐπὶ τ]
[ῆς Πανδι]ονίδος δεκάτης [πρυτα]
[νεία]ς ἧι Εὐθυγένης Ἡφαισ[τοδή]
[μο]υ Κ[ηφ]ισιεὺς ἐγρα[μ]μάτε[νεν· Θ]
5  [αρ]γηλιῶνος ἔνηι καὶ ν[έ]αι, [ἑβδό]
[μηι κ]αὶ τριακοστῆι [τ]ῆ[ς] πρ[υταν]
[είας· –] – – – – κτλ. – – – – – –

In an intercalary year the last prytany should have normally 38 days. One may assume that Skirophorion ἔνη καὶ νέα was followed by ἔνη καὶ νέα ἐμβόλιμος, being respectively the 29th and 30th days of the month. Skirophorion 29 was equated with

---

[56] Cf. Pritchett and Neugebauer, *Calendars*, p. 60.

Prytany X 37, and the equation is written as follows:

⟨Skirophorion⟩ 29 = Prytany X 3 [7] = 383rd day.

Months and prytanies were arranged as follows (possibly):

Months    29 30 29 30 29 30 29 30 29 30 29 30 30 = 384
Prytanies  39 39 39 39 38 38 38 38 38 38 = 384.

Counting back from Skirophorion 29, one finds that 57 days had elapsed since Thargelion 2. If the ninth prytany, like the tenth, had 38 days, Thargelion 2 should have been equated with Prytany IX 18. It seems to me no more difficult to assume an error from this norm than any other. Instead of holding that 23 was mistakenly written for 13,[57] possibly ΔΔΙΙΙ read incorrectly for ΔΙΙΙ, I would suppose that 23 was mistakenly written for 18, ΔΔΙΙΙ, possibly, for ΔΓΙΙΙ, and I should give the equation in *I.G.*, II², 373, as:

Thargelion 2 = Prytany IX ⟨18⟩ = 326th day.

### 321/0 (Ordinary)
### Archon Archippos
### *I.G.*, II², 385

No equation is preserved. For the ἀναγραφεῖς see Pritchett and Meritt, *Chronology*, pp. xvi-xvii; Meritt, *Hesperia*, XIII, 1944, p. 235; XXVI, 1957, p. 53.

*I.G.*, II², 546 (cf. Pritchett and Meritt, *Chronology*, p. 6)

a. 321/0 a.

[ἀναγραφεὺς — — — — — — — — — — — ἐξ Οἴου]
[ἐπὶ Ἀρχίππ]ου ἄρχοντος ἐπὶ τῆ[ς ....¹¹.....]    ΣΤΟΙΧ. 36
[...⁶... πρυτ]ανείας ἧι Ἀριστ[......¹³......]

---

[57] One of the suggestions that has been put forward on the supposition that the year was ordinary. Cf. Pritchett and Neugebauer, *Calendars*, p. 60.

[.... ἐγραμμ]άτενε· δεκάτει ὑ[στέραι ....⁷...]
5  [.....¹¹.....] καὶ δεκάτει τῆ[ς πρυτανείας· τῶ]
κτλ.

There is no evidence here for the calendar character of the year, but it was undoubtedly ordinary, since the years immediately before and after were both intercalary.

320/19 (Intercalary)

Archon Neaichmos

*I.G.*, II², 380

*a.* 320/19 *a.*

Ἀναγραφεὺς Ἀρχέδ[ι]κος Ναυκρίτου Λαμπτ[ρεύ]ς
Ἐπὶ Νεαίχμου ἄρχοντος ἐπὶ τῆς Ἐρεχθη        ΣΤΟΙΧ. 31
ἱδος δευτέρας πρυτανείας εἶ Θηρα[μ]έν
ης Κηφισιεὺς ἐγρα[μμ]άτενε· Βοηδρ[ομ]ιῶ
5  νος ἐνδεκ[ά]τει, [μ]ιᾶι καὶ τ[ρ]ιακοστεῖ τῆ
ς πρυτ[α]νείας·  ‒ ‒ ‒ ‒ κτλ. ‒ ‒ ‒ ‒ ‒ ‒

Boedromion 11 = Prytany II 31 = 70th day.

The year was intercalary, with 59 days in the first two months, and with 39 days in the first prytany.

*I.G.*, II², 383*b* (*Addenda*: EM 12456)

*a.* 320/19 *a.*

[Ἀναγ]ραφεὺ[ς Ἀρχέδικος Ναυκρίτου Λαμπτρεύς]
       θ        ε        [ο        ί]
    ἐπὶ Νεαίχμ[ου ἄρχοντος ἐπὶ τῆς]        ΣΤΟΙΧ. 25
    Πανδιονίδ[ος τρίτης ᵛ πρυτανε]
5  ίας ἦι Στρατ[ωνίδης Παιανιεύς]
    ἐγραμμάτευ[εν· δευτέραι φθίνο]
    ντος, δε[κ]άτη[ι τῆς πρυτανείας· ἐ]
    κκλησία· τῶν π[ροέδρων ἐπεψήφι]

114 The Fourth Century

ζεν Τελέσιππ[ος ....ˢ....· ἔδοξ]
10  εν τεῖ βουλε[ῖ καὶ τῶι δήμωι· Δημ]
άδης Δημέ[ου Παιανιεὺς εἶπεν· ἐ]
πειδὴ οἱ [ – – – – – – – – – – –]
⟨ΓΟΥ[– – – – – – – – – – – – – – –]

⟨Boedromion⟩ 2[9] = Prytany [III] 10 = 88th day.

The text is based upon a study of the stone made in Athens in 1958.

The calendar equation has been given recently by Pritchett and Neugebauer (*Calendars*, p. 62) as

⟨Thargelion⟩ 2[9] = Prytany [X] 10 = 356th day.

But this equation brings an element of irregularity into any calendar scheme, which they resolve by assuming manipulation of the festival calendar: "From the equation in *I.G.*, II², 383*b*, we conclude that one or two extra days were intercalated in Mounichion or Thargelion with compensatory shortening of the end of the final month, Skirophorion. The inscriptions are so widely dispersed throughout this year that we can conjecture that it was intercalary with 385 days in which six of the first ten months were full and four hollow followed by 29 + 2, 30, and 28." Klaffenbach queries whether it may not have been a year of 384 days with Skirophorion having only 27.[58]

There is no justification in the lunar calendar for assuming here an intercalary year which had 385 days. Parker and Dubberstein[59] give a table of dates of computed first visibility of the lunar crescent at Babylon, which I have transcribed, and to which I have added the names of the Athenian months:

---

[58] *Gnomon*, XXI, 1949, p. 135.
[59] *Babylonian Chronology*, p. 36.

| July 9 | Hekatombaion | 29 |
|--------|--------------|----|
| August 7 | Metageitnion | 30 |
| September 6 | Boedromion | 30 |
| October 6 | Pyanepsion | 30 |
| November 5 | Maimakterion | 29 |
| December 4 | Posideon | 30 |
| January 3 | Posideon II | 29 |
| February 1 | Gamelion | 30 |
| March 3 | Anthesterion | 29 |
| April 1 | Elaphebolion | 30 |
| May 1 | Mounichion | 29 |
| May 30 | Thargelion | 29 |
| June 28 | Skirophorion | 30 |
| July 28 | | |

Babylon has a longitude and latitude, of course, different from those of Athens, but the observations (if made) ought to have been nearly enough the same in Athens to show that the length of the year was probably a normal 384 days.

On the other hand, I agree with Pritchett and Neugebauer that Kirchner's arrangement of prytanies, in which the ninth prytany is supposed to have 37 days and the tenth prytany 40 days, is not satisfactory, if something better can be found.[60]

---

[60] This was my arrangement when I wrote in *Hesperia*, XIII, 1944, pp. 236-241.

The question is, I think, how can the year be interpreted with least irregularity to any calendar scheme. I hold that any combination of prytanies of 38 and 39 days that will add up to 384 is not irregular; also that any alternation of full and hollow months that does not have three full or two hollow months together is not irregular.

We may, then, see how the interposing of an extra day (or more, if need be) affects the problem. First of all, it will have made the end of Mounichion or Thargelion badly out with the moon, especially since Pritchett and Neugebauer have already assumed that six of the first ten months were full. I believe also that a maladjustment deliberately introduced should have been corrected soon after the occasion for it was past, and in principle do not believe that extra days added in Mounichion were corrected only at the end of the month of Skirophorion.[61] One expects, for example, that any intercalation in Thargelion to postpone the celebration of the Plynteria will have been compensated well before Θαργηλιῶνος δευτέρα φθίνοντος.[62] When that day was reached, it was too late to make a correction in Thargelion.

The irregularity, if not in the calendar, may have been in the record, and this approach to a solution must be studied. The date by prytany in *I.G.*, II², 383*b*, was surely the tenth. At least, this is the numeral inscribed on the stone, and it may be seen clearly in the photograph published by Leonardos in ᾿Αρχ. Δελτίον, I, 1915, p. 195. But the ordinal number of the prytany is not preserved; it must be restored in line 4. In the stoichedon text πέμπτης, ἑβδόμης, and δεκάτης satisfy exactly the requirements of the space available. We have just been discussing the difficulties of δεκάτης. The restoration of πέμπτης is forbidden by the fact that *I.G.*, II², 381, shows Antiochis (not Pandionis)

---

[61] See above, pp. 103-104.
[62] See above, p. 104, for possibly such a case.

to have held the fifth prytany. If one restores ἑβδόμης the anomaly with the calendar is greater than with δεκάτης.

If the ordinal numeral with the prytany was τρίτης the year can be kept normal, but the irregularity must be attributed to the record. This seems to me a lesser evil. The year of the prytanies remains normal, the year of the months remains normal, and the record as restored is in harmony with and correct for both calendars. The cost has been one uninscribed space assumed in line 4.

The occasional uninscribed space in an inscription may be quite inexplicable on any attempt at rationalization. Why, for example, were six spaces left uninscribed in the phrase ἐπὶ τῆς Οἰνεῖδος *vv[vvvv]* ἐ[β]δόμης π[ρυτα]νείας of *I.G.*, II², 910? Lattermann thought he saw traces that showed the word ἑβδόμης to have been cut twice and once erased (cf. Kirchner's note *ad loc.*). But Sterling Dow affirms [63] that the alleged erasure is too short for ἑβδόμης and, moreover, that there were at least two spaces after Οἰνεῖδος that had never been inscribed. He thinks that space may have been left for the numeral to be inserted after the first cutting of the text, and that too much allowance was made. Perhaps so; and perhaps space was left here in line 4 that was not completely filled by τρίτης. Or the anomaly of this line may be similar to that at the end of the second line of *I.G.*, II², 704. Here technical difficulties at the end of the line left only 26 letters in a text that was otherwise consistently and strictly stoichedon.[64] It is more difficult to account, even hypothetically, for the uninscribed space between [Μυ]ρρινούσι[ος] and ἐγ[ρ]αμμ[ά]τευεν in *I.G.*, II², 768, yet the space, epigraphically, must be assumed (the text is stoichedon) and the line must be read [Μυ]ρρινούσι[ος*ᵛ*] ἐγ[ρ]αμμ[ά]τευεν.[65]

[63] *Hesperia*, Suppl. I, p. 133.
[64] Reference should be made to *Hesperia*, XXVI, 1957, p. 56 note 8, for a full account.
[65] Cf. *Hesperia*, VII, 1938, p. 144. For a suggested explanation of the two

*I.G.*, II², 381

*a.* 320/19 *a.*                    ΣΤΟΙΧ. 26

Ἀναγραφεὺς Ἀ[ρχέδικος Ν]αυκρ[ίτ]
[ο]υ Λαμπτρεύς· [ἐπὶ Νεαί]χμου ἄρχ[ο]
[ν]τος ἐπὶ τῆς Ἀν[τιοχ]ίδος πένπτ[η]
[s π]ρυτανείας ἧ[ι Νι]κόδημος Ἀναφ
5   λύ[σ]τιος ἐγρα[μμάτ]ευε[ν]· Ποσιδεῶ
νος ὑστέρου τ[ετρά]δι ἐπὶ δέκα, ἔκ
[τ]ηι καὶ τριακ[οστῆι] τ[ῆ]ς πρυτανε
[ί]ας· — — — — — κτλ. — — — — — —

Posideon II 14 = Prytany V 36 = 192nd day.

With alternation of full and hollow months 30 29 30 30 29 30 and with prytanies 39 39 39 39 the fourteenth day of second Posideon is the same as the thirty-six day of the fifth prytany.

*I.G.*, II², 382

*a.* 320/19 *a.*                    ΣΤΟΙΧ. 31

[θ]            ε          ο            [ι]
[Ἀναγραφεὺς Ἀρχέδ]ικος Ναυκρίτο[υ Λαμ]
[πτρεύς· ἐπὶ Νεαίχ]μου ἄρχοντος ἐπ[ὶ τῆς]
[Ἀντιοχίδος πέμπτ]ης πρυτανείας [ἧι Νι]
5   [κόδημος Ἀναφλύστι]ος ἐγρ[αμ]μάτε[νεν· Π]
[οσιδεῶνος ὑστέρου τ]ετράδι ἐπὶ δέκ[α, ἔ]
[κτηι καὶ τριακοστῆι τ]ῆς πρυτανεία[s· —]
κτλ.

This decree was passed on the same day with *I.G.*, II², 381.

uninscribed spaces in the second line of the text of *Hesperia*, VIII, 1939, p. 31, which have been accepted by Pritchett and Neugebauer (*Calendars*, p. 65) and where the blanks probably came either just before or just after the numeral, see below, p. 126.

*I.G.*, II², 336*b* (cf. Pritchett and Meritt,
*Chronology*, p. 7)

*a.* 320/19 *a.*

5    ['Ελαφ]ηβολιῶνος ἔνε[ι καὶ νέαι, ἕκτει κα]
      [ὶ εἰκ]οστεῖ τῆς πρυτ[ανείας – – κτλ. – –]

Elaphebolion 30 = Prytany ⟨VIII⟩ 2[6] = 296th day.

With alternation of full and hollow months 30 29 30 30 29
30 29 30 29 30 and with prytanies 39 39 39 39 38 38 38 the
last day of Elaphebolion (30th) is the same as the 26th day of
the eighth prytany.

*Hesperia*, XIII, 1944, pp. 234-235 (6)

*a.* 320/19 *a.*

   [..κόστρατος – – –]λωνος    Φιλ[ιππεύς?]
                                ΣΤΟΙΧ. 38
   ['Επὶ Νεαίχμου ἄρχοντος, ἐπ'] ἀναγραφέω[s δὲ 'Αρχεᵛ]
   [δίκου τοῦ Ναυκρίτου Λαμπ]τρέως ἐπὶ τῆ[s Οἰνεῖᵛ]
   [δos ὀγδόης πρυτανείας ἧι ...]νων 'Οῆθ ἐγραμ[μάτᵛ]
5    [ευεν· Μουνιχιῶνος ὀγδόηι ἰσ]ταμένου, τετ[άρτηι]
   [καὶ τριακοστῆι τῆς πρυτανεί]ας· ἐκκλησί[α κυρᵛ]
   [ία· τῶν προέδρων ἐπεψήφιζεν 'Ι]οφῶν Στει[ρ καὶ σᵛ]
   [υμπρόεδροι· ἔδοξεν τῶι δήμωι·] Δημάδης Δη[μέουᵛ]
   [Παιανιεὺς εἶπεν· ᵛᵛ ἐπειδὴ ..]κόστρατο[s ...⁵..]
10    [– – – – – – – – – – – – – – – –]τε τῶν ἐς Σ[...⁶...]
   [– – – – – – – – – – – – – – – –] 'Αθηναίων μη[....]
   [– – – – – – – – – – – – – – – –] τοὺς ἐπιβουλ[...]
   [– – – – – – – – – – – – – – – –] 'Αθηναι[...⁵..]
   [– – – – – – – – – – – – γυναι]ξὶ καὶ παι[σὶᵛ]
15    [– – – – – – – – – – – – – – ]Ε. ὁ μέλλων Λ[...]
   [– – – – – – – – – – – – – – κα]ταληφθῆναι [...]
   [– – – – – – – – – – – – – –]ων καὶ διΛ[...⁵..]

[— — — — — — — — — — — — — — — — —]ν αὐτὸν ᾿Αθ[ηνα<sup>υ</sup>]

[ἰου — — — — — — — — — — — —? ἐπίστατ]αι ὁ δῆμος ὁ [᾿Αθ<sup>υ</sup>]

20  [ηναίων — — — — — — — — — — — — — —]ο πλέον τ[ . . . . ]

[ — — — — — — — — — — — — — — — — —]ι μετ[ . . . . ]

[— — — — — — — — — — — — — — — — —]τα [ . . .<sup>5</sup>. . ]

[— — — — — — — — — — — — — — — — — —]ι[ . . .<sup>5</sup>. . ]

[Mounichion 8] = Prytany [VIII] [3]4 = 304th day.

The decree was passed eight days after that of *I.G.*, II², 336*b*.
Since publishing this text in *Hesperia* (*loc. cit.*) I have been
able to study the stone in the Museum of the Agora at Athens
and to determine the right margin.[66]  In line 2 there was room
for eight letters after the omega of ἀναγραφέω[ς]; there is, of
course, no assurance that all these spaces were inscribed, for
there may have been a margin of at least one letter-space left
free. I have assumed here that this was the case, and the new
text is based on a stoichedon line of 38 letter-spaces, 37 of them
regularly occupied by the stoichedon pattern and the extra
space once used (in line 5) to accommodate the final iota of
τετ[άρτηι]. But I believe that this concession does not argue for
syllabic division elsewhere except as it fell in accidentally with
the stoichedon pattern.

Other inscriptions of the year of Neaichmos (*I.G.*, II², 383
and 384) yield no calendar equations. The months and prytanies
may now be ordered as follows:

Months    30  29  30  30  29  30  29  30  29  30  29  30  29 = 384
Prytanies  39    39    39    39    38    38    38    38    38    38 = 384.

There are no observable calendar irregularities in this year.

---

[66] See *Hesperia*, XIII, 1944, pp. 238-240; Pritchett and Neugebauer, *Calendars*,
p. 62.

319/8 (Ordinary)

Archon Apollodoros

This year was the subject of a special study made by Pritchett[67] before the writing of the *Calendars of Athens*. In 1947 Pritchett and Neugebauer claimed that the festival year as reconstructed earlier by Pritchett was too rigid.[68] This is, I think, questionable. The prytanies throughout the year followed Aristotle's rule for lengths of prytanies and the months alternated full and hollow. All was in order, with no irregularity except the name of a month miswritten in *I.G.*, II², 388. When Pritchett and Neugebauer restored *I.G.*, II², 388, in such a way that {Mounichion} [12] might fall on Prytany VIII 2[8], they implied, either consciously or unconsciously, a festival year in which the alternation of full and hollow months was reversed, the tangible difference being that they must have assumed Elaphebolion to be hollow rather than full. The order of alternation makes very little difference except as it abuts the end of the preceding year and the beginning of the following year, and here, I think, Pritchett's original arrangement was the better.

*Hesperia*, IX, 1940, p. 345 [69]

a. 319/8 a.

θ ε ο ί

Πολιτεία Αἰνήτωι Ῥοδίωι

Ἀναγραφεὺς Εὔκαδμος Ἀνακαιεύς <sup>υ</sup>      ΣΤΟΙΧ. 28

ἐπὶ Ἀπολλοδώρου ἄρχοντος ἐπὶ τῆς

---

[67] *Hesperia*, X, 1941, p. 269.

[68] *Calendars*, p. 63 note 56.

[69] In this inscription no letters were crowded at the beginning of line 7 (see Pritchett and Neugebauer, *Calendars*, p. 63, with [ϵἰκοσ]τϵῖ, which they have taken over from Schweigert's text in *Hesperia*, IX, 1940, p. 345). Rather, the word καί was complete at the end of line 6, with some loss of the true stoichedon pattern. See the photograph in *Hesperia*, IX, 1940, p. 347.

5 [. . .]ντίδος τετάρτης πρυτανείας· Μ
[αιμακ]τηριῶνος ἐνδεκάτει, μιᾶι καὶ
[εἰκοσ]τεῖ τῆς πρυτανείας· – κτλ. –

Maimakterion 11 = Prytany IV 21 = 129th day.

*I.G.*, II², 386 (= Wilhelm, *Abh. Ak. Berlin*, Jahrgang 1939, no. 22, pp. 22-23; Pritchett and Neugebauer, *Calendars*, p. 63).

*a.* 319/8 *a.*

θ ε ο ί
Πολιτεία Ἀμυντ[. . . . .⁹. . . .]
Ἐ[π]ὶ Ἀπολλοδώρου ἄ[ρχοντος ἐπὶ τῆς Α]  ΣΤΟΙΧ. 29
[ἰγη]ίδος ἕκτης π[ρυτανείας· γραμματ]
5 [εὺς] Δωσί[θεος – – – – κτλ. – – – –]

There is no calendar equation.

*Hesperia*, VII, 1938, pp. 476-479 (31)

*a.* 319/8 *a.*                  ΣΤΟΙΧ. 26

[Ἐπὶ Ἀπο]λλοδ[ώρ]ου ἄρχοντος ἐπ[ὶ τ]
[ῆς Ἀντι]οχί[δος ἑ]βδόμης πρυτα[νε]
[ίας καὶ ἀναγραφέ]ως Εὐκάδμου Ἀ[ν]
[ακαι]έως· Ἐλαφ[ηβο]λιῶνος δωδεκ[ά]
5 τει, τετάρτει [καὶ τ]ριακοστεῖ τ[ῆ]
ς πρυτανείας· – – – – κτλ. – – – –

Elaphebolion 12 = Prytany VII 34 = 248th day.

*I.G.*, II², 388 (cf. *Hesperia*, VII, 1938, p. 479)

*a.* 319/8 *a.*                  ΣΤΟΙΧ. 29

[Ἐ]πὶ Ἀπολλοδώρου [ἄρχοντος καὶ ἀναγ]
[ρ]α[φέ]ως Εὐκάδμου [Ἀνακαιέως ἐπὶ τῆς]
[Ἐρεχ]θεῖδος ὀγδό[ης πρυτανείας εἶ Φ]
[ιλ]οκτήμων Κηφισ[ιεὺς ἐγραμμάτευε]
5 [ν·] Ἐλαφηβολιῶνο[ς δωδεκάτει, ἐνάτει]

καὶ εἰκοστεῖ τῆ[ς πρυτανείας· – κτλ. –]

⟨Mounichion⟩ [12] = Prytany VIII 2[9] = 278th day.[70]

Pritchett and Neugebauer, *Calendars*, p. 64, note that "there is clearly some error in the calendar equations of this year and we have followed the suggestion of M. Crosby (*Hesperia*, VII, 1938, p. 479), adopted by W. B. Dinsmoor (*Athenian Archon List*, p. 34, note 62) and W. K. Pritchett (*Hesperia*, X, 1941, p. 269), that the name of the civil month in *I.G.*, II², 388 was inscribed by error as Elaphebolion instead of Mounichion."

I call attention to an instance noted above (pp. 111-112) in which the name of a month was wrongly inscribed.

*I.G.*, II², 387 (cf. *Hesperia*, VII, 1938, p. 478)

a. 319/8 a.                    NON-ΣTOIX.

['Επὶ 'Απολλοδώ]ρου ἄ[ρχοντο]
[ς ἐπὶ τῆς δεκά]της π[ρυτανε]
[ίας· γραμματεὺ]ς 'Αφόβητο[ς Ko]
[θωκίδης – – – – ] – κτλ. – – –

There is no calendar equation, but the demotic of the secretary shows that the tenth prytany belonged to Oineis.

*I.G.*, II², 390 (cf. Dinsmoor, *Archons*, p. 22)

a. 319/8 a.                    ΣTOIX. 33

[θ]            ε            [ο            ί]
'Επὶ ἀναγρα[φέως Εὐκάδμου 'Ανακαιέως καὶ]
['Απ]ολλοδώρ[ου ἄρχοντος, ἐπὶ τῆς Οἰνηίδος]
[δε]κάτης πρ[υτανείας ⁿ Σκιροφοριῶνος τε]
5    [τρά]δι [– – – – – – – – – – – – – – – –]

There is no calendar equation.

[70] The equation of Mounichion 11 with Prytany VIII 28 as the 277th day would be equally satisfactory.

Hesperia, X, 1941, p. 268 (69)

a. 319/8 a. ΣΤΟΙΧ. 30

.['Επ' 'Απολλοδώρου ἄρχοντος, ἀναγραφέω]
[ς δὲ Εὐκάδμ]ου 'Αν[ακ]αέω[ς, ἐπὶ τῆς Οἰνεῖ]
[δος δεκάτ]ης πρυτανεία[ς· Σκιροφοριῶ]
[νος ἔνε]ι καὶ νέαι, πένπτ[ει καὶ τριακο]
5 [στεῖ τῆ]ς πρυτανείας· — — [ — — κτλ. — ]

[Skirophorion] 29 = Prytany [X] [3]5 = 354th day.

The months and prytanies of the year may now be arranged
as follows, on the evidence of the foregoing inscriptions:

Months    30  29  30  29  30  29  30  29  30  29  30  30 = 355
Prytanies 36  36  36  36  35  35  35  35  35  36 = 355.

I have indicated a year of 355 days, in order to have the first
day of the following year fall on July 18, though I do not hold
that this is necessary in the Athenian calendar. The year, at
Babylon, and timed from the observable crescent moon at
beginning and end, comprised the days from July 28, 319 B.C.
through July 17, 318 B.C., a span of 355 days.[71]

The text of I.G., II², 388, can be retained, as inscribed on the
stone, if one assumes the intercalation of extra days in the festival
calendar before the end of Elaphebolion, but after Elaphebolion
11 (or 12). In this case, the equation of I.G., II², 388, will read
as in the Corpus: 'Ελαφηβολιῶνο[ς ἔνει καὶ νέαι, μιᾶι] καὶ εἰκοστεῖ
τῆ[ς πρυτανείας]. Actually, twenty-two days elapsed between
Prytany VII 34 of the preceding equation and Prytany VIII 21;
hence, in the festival calendar, one must assume the intercalation
of four ἡμέραι ἐμβόλιμοι to make the span from Elaphebolion 12
to Elaphebolion 30 amount also to twenty-two days.

There seems to be here something of the same set of circum-

[71] See Parker and Dubberstein, Babylonian Chronology, p. 36.

stances that existed in the archonship of Pytharatos (271/0 B.C.) when four days were added to the festival calendar, also in the month of Elaphebolion.[72] Pritchett has emphasized quite recently the importance of this text from the year of Pytharatos in the study of the Athenian calendar,[73] claiming even that it vindicates the thesis "that the festival calendar was made up of months of uneven duration." While it confirms the thesis that the archons could and did postpone dates in the festival calendar by the intercalation of days, it does not, I think, go so far as to show a festival calendar "made up of months of uneven duration." An irregularity of this kind, now and then, in 271/0 and possibly in 319/8, affecting in each instance only one (or possibly two) months in the year must be weighed against the evidence of the other texts—especially those of the fourth century—for a most extraordinary uniformity in the progression of the festival calendar.

Pritchett and Neugebauer have not claimed the evidence of *I. G.*, II², 388, in support of their advocacy of the tampered festival year. Nor do I. Had the extra days been needed to postpone the celebration of the Dionysia, and had the compensation been possible before the end of the month, I should have done so. As it is, I believe that there was no irregularity in the festival calendar of this year, only an error in the record.

### 318/7 (Ordinary)
### Archon Archippos

The inscriptions of this year have offered a number of perplexing problems. Pritchett and Neugebauer are undoubtedly right in claiming the year as an ordinary year. I have again examined in Athens the four texts which give the calendar equations, and offer now the following readings and restorations.

[72] See the text in *Hesperia*, XXIII, 1954, p. 299, lines 3-4: Ἐλαφηβολιῶνο[ς ἐ]νάτει ἱσταμένου τετάρτει ἐμβολίμωι.
[73] *B.C.H.*, LXXXI, 1957, pp. 274-275.

I. G., II², 448 (lines 35-38)

a. 318/7 a.                                    ΣΤΟΙΧ. 41

35  Ἐπὶ Ἀρχίππου ἄρχοντος ἐπὶ τῆς Ἀκ[αμαντίδος τετάρ]
    της πρυτανείας ἧι Θέρσιππος Ἱππο[...⁶... Κολλυτε]
    ὺς ἐγραμμάτευε· Μαιμακτηριῶνος ἕ[κτει ⟨μετ'⟩ εἰκάδας,
                                                        πέμ]
    πτει καὶ τριακοστεῖ τῆς πρυτανεία[ς· τῶν προέδρων]
    κτλ.

Maimakterion [25] = Prytany IV 35 = 143rd day.

For commentary on this text, see above, pp. 55-58.

Hesperia, VIII, 1939, pp. 31-32

a. 318/7 a.                                    ΣΤΟΙΧ. 23
    Ἐπὶ Ἀρχίππ[ου ἄρχοντος ἐπὶ τ]
    ῆς Κεκροπίδ[ος ἕκτης ˇˇ πρυταν]
    έας ἧι Θ[έ]ρ[σιππος Ἱππο...⁵..]
    [. Κ]ολλυ[τ]ε[ὺς ἐγραμμάτευε· Γα]
5   μηλι[ῶνο]ς [ὀγδόει ἐπὶ δέκα, ἕκ]
    [τ]ει καὶ δε[κάτει τῆς πρυτανε]
    [ίας· ἐκ]κλ[ησία – – κτλ. – – ]

Gamelion [18] = Prytany [VI] 1[6] = 195th day.

In the ninth space in line 5, my examination of the stone leads me to believe that there is no trace of any preserved letter, and that epigraphically any letter is a possible restoration.[74]

The corruption of the stoichedon order in line 2 was caused, apparently, by the cutting of Κροπίδος instead of Κεκροπίδος. Initial K is in its proper stoichos, and Π is in its proper stoichos; the intervening PO were erased and replaced by EKPO. Some of the letters of the third and later lines may also have been

[74] See also the note in Pritchett and Neugebauer, Calendars, p. 65.

cut, or at least blocked out, before the error in line 2 was detected. This will explain the troublesome two uninscribed spaces in line 2. They would not have been uninscribed if Κεκροπίδος had been correctly spelled in the first place.

*Hesperia*, IV, 1935, p. 35

a. 318/7 a. ΣΤΟΙΧ. 89

Ἐπὶ Ἀρχίππου ἄρχοντος [ἐπὶ τῆς Κεκροπίδος ἕκτης πρυτανείας
ἧι Θέρσιππος Ἱππο . . .ˢ. . . Κολλυτεὺς ἐγραμμά<sup>vv</sup>]
τευεν· Γαμηλιῶνος ἔνει [καὶ νέαι, ὀγδόει καὶ εἰκοστεῖ τῆς πρυ-
τανέας· ἐκκλησία κυρία· τῶν προέδρων ἐπεψήφιζεν] ⁷⁵

Gamelion 30 = Prytany [VI 28] = 207th day.

*I.G.*, II², 350 (cf. *Hesperia*, VIII, 1939, p. 33)

a. 318/7 a. ΣΤΟΙΧ. 26

['Επὶ Ἀρχίππου ἄρχοντος ἐπὶ τῆς . ]
[. . .ˢ. . . ἑβδόμης πρυτανεί]α[ς ἧι Θ]
[έρσιππος Ἱππο . . .ˢ. . . Κ]ολλυτε[ὺς ἐγρ]
[αμμάτευε· Ἀνθεστ]ηριῶνος ἑ[βδόμ]
5 [ει φθίνοντος, ἕκτ]ει καὶ δεκάτ[ει]
[τῆς πρυτανείας· ἐ]κκλησία ἐν Δι[ο]
κτλ.

Anthesterion [23] = Prytany [VII] 1[6] = 230th day.

With Anthesterion a hollow month, the date ἑ[βδόμει φθίνον-
τος] becomes the 23rd day (see above, p. 45). Months and prytanies within the year may now be arranged as follows:

Months    29  30  29  30  30  29  30  29  30  29  30  29 = 354
Prytanies 36  36  36  36  35  35  35  35  35  35 = 354.

There is no irregularity.

⁷⁵ I have followed Pritchett and Neugebauer, *Calendars*, p. 65, in positing a stoichedon line of 89 letters and in assuming syllabic division.

　　　　　The Fourth Century

317/6 (No Evidence)
Archon Demogenes

316/5 (No Evidence)
Archon Demokleides

315/4 (No Evidence)
Archon Praxiboulos

314/3 (Intercalary)
Archon Nikodoros

*I.G.*, II², 450

*a.* 314/3 *a.*　　ΣΤΟΙΧ. 21 (Irregular)

Ἐπὶ Νικοδώρου ἄρχοντος ᵛ
ἐπὶ τῆς Κεκροπίδος ἕκτη
ς πρυτανείας· Γαμηλιῶνος
ἑνδεκάτηι, ἕκτηι καὶ εἰκο
5　στῆι τῆς πρυτανείας· – κτλ. –

Gamelion 11 = Prytany VI 26 = 218th day.

Pritchett and Neugebauer bring the equation to the 220th day of the year, and write:[76] "In order for Gamelion (VII) 11 to fall on the 220th day of a normal intercalary year we would have to assume that six of the first seven months were of 30 days. This is astronomically impossible, so we assume that the archons had intercalated approximately two extra days in the first part of the year." But the fixing of the day as the 220th rests itself on an assumption, namely, that the prytanies were always—in each intercalary year—four 39's followed by six 38's, a rule derived by inference only (rigidly applied) from Aristotle's account of the ordinary year with quite different prytanies.

[76] *Calendars*, p. 66.

If Gamelion 11 was the 218th or 217th day of the year, the following arrangements of months and prytanies are possible:

### I

| Months | 30 | 29 | 30 | 29 | 30 | 29 | 30 | 29 | 30 | 29 | 30 | 29 | 30 = 384 |
|---|---|---|---|---|---|---|---|---|---|---|---|---|---|
| Prytanies | 39 | 39 | 38 | 38 | 38 | 38 | 38 | 38 | 39 | 39 | | | = 384 |

### II

| Months | 29 | 30 | 29 | 30 | 29 | 30 | 29 | 30 | 30 | 29 | 30 | 29 | 30 = 384 |
|---|---|---|---|---|---|---|---|---|---|---|---|---|---|
| Prytanies | 39 | 38 | 38 | 38 | 38 | 38 | 38 | 39 | 39 | 39 | | | = 384. |

There are numerous insignificant ways of ordering the prytanies differently, but our study so far has shown that neither of these schemes should be considered irregular.

313/2 (No Evidence)
Archon Theophrastos

*I.G.*, II², 451

312/1 (No Evidence)
Archon Polemon

311/0 (No Evidence)
Archon Simonides

310/09 (Ordinary)
Archon Hieromnemon

*I.G.*, II², 453

*a.* 310/09 *a.*          ΣΤΟΙΧ. 34

Ἀ ν τ [ – – – – – ]
['E]πὶ ['Ιερο]μνή[μονος ἄρχοντος ἐπὶ τῆς ...ντ]
[ί]δ[ο]ς ἕκτης π[ρυτανείας· Γαμηλιῶνος ἐνάτη]
[ι ἐπὶ] δέκα, ἐν[άτηι καὶ δεκάτηι τῆς πρυτανε]
5   [ία]ς· – – – – – – – κτλ. – – – – – – – – –

[Gamelion] 1[9] = Prytany VI [1]9 = 196th day.

Not holding to a rigid conciliar year, I see no advantage in substituting new restorations for those given in the *Corpus*. For the year, see the commentary on *I.G.*, II², 453. The months and prytanies may be arranged as follows:

Months   30  29  30  29  30  29  30  29  30  29  30  29 = 354 [77]
Prytanies 36  36  35  35  35  35  35  35  36  36 = 354.

## Unknown Year (Intercalary)

### *I.G.*, II² 449

ΣTOIX. 24

['E]πὶ τῆς Αἰαντίδ[ος ἕκτης πρυτ]
[α]νείας· ὀγδόηι [ἐπὶ δέκα, ἕκτηι]
τῆς πρυτανείας· [— — κτλ. — — —]

⟨Posideon II⟩ [1]8 = Prytany [VI 6] = 196th day.

The year is intercalary, with months and prytanies disposed as follows: [78]

Months   30  30  29  30  29  30  29  30  29  30  29  30  29 = 384
Prytanies 38  38  38  38  38  38  39  39  39  39 = 384.

There is no irregularity. In holding to the assumption that in an intercalary year the first four prytanies must have had 39 days each, Pritchett and Neugebauer suggest that two letters were cut in place of one somewhere in the ordinal numeral of the phyle, making the prytany either eighth or ninth; or, assuming an ordinary year in which the first four prytanies had 36 days each, they suggest that ὀγδόηι ἐπὶ δέκα was cut on the stone though the date by month should have been ὀγδόηι ἱσταμένου.[79]

They might also have avoided a calendar equation altogether

---

[77] Or in reverse alternation.
[78] See above, p. 100, for a reverse order of prytanies in 327/6. See also above, p. 99 note 37.
[79] *Calendars*, p. 67.

by assuming that two letters were cut in place of one in the
date by prytany and reading ὀγδόηι [καὶ τριακοστῆι] τῆς
πρυτανείας.

In the preceding pages we have reviewed the known calendar
equations from 346 to 307 B.C. Those of significance have num-
bered 61 (or 62), distributed as follows: 346/5 (one), 341/0
(one), 338/7 (one), 337/6 (three), 336/5 (three), 335/4
(three), 334/3 (one), 333/2 (four), 332/1 (four), 331/0
(three), 330/29 (two), 329/8 (two), 328/7 (two), 327/6 (two,
possibly three), 326/5 (one), 325/4 (two), 324/3 (three),
323/2 (four), 322/1 (three), 321/0 (none), 320/19 (five),
319/8 (four), 318/7 (four), 314/3 (one), 310/09 (one), Unknown Year (one).

Of these 61 (or 62) equations only two have indicated irregularities in the festival calendar caused by abnormal intercalation
of extra days. In one instance (325/4) the intercalation was
of three days, and in one instance (333/2) of one day. In both
cases the correction was effected, so far as the evidence shows,
promptly. There has been so far no extreme example such as
Pritchett and Neugebauer emphasize from the evidence of the
second century.[80] It seems rather that one must assume less
irregularity in the fourth century than in the second, a consideration which has already been raised by Gomme,[81] and that
merely as a matter of averages there is a very good chance that
epigraphical festival dates in the fourth century, as we find them,
will be κατὰ θεόν as well as κατ᾽ ἄρχοντα.[82]

[80] See now Pritchett, B.C.H., LXXXI, 1957, p. 280: "If the archon of 145/4
B.C. could intercalate 20 days, we must conceive of the Athenian festival
calendar as being far removed from the lunar movement. We must conclude,
also, in the absence of contrary evidence, that if the archon of 145/4 B.C. had
this power, the archons of other years probably also possessed a similar right."
[81] A. W. Gomme, Commentary on Thucydides, III (1956), p. 714 note 3.
[82] An extreme example of retardation in the festival calendar, not at Athens
in the fourth century but at Argos in the fifth century, is in Thucydides, V,
54, 3: Ἀργεῖοι δ᾽ ἀναχωρησάντων αὐτῶν τοῦ πρὸ τοῦ Καρνείου μηνὸς ἐξελθόντες τετράδι

It is also quite clear from the epigraphical record that the sequence of intercalary and ordinary years is known approximately from 346/5 to 310/09, and absolutely from 337/6 to 319/8,[83] as well as for 318/7. This being so, one can give Julian dates to the known decrees of these certain years, with confidence that within a day or two in any given instance the date will be correct. The following dates are now available for the years from 337/6 to 318/7, fixing the beginnings and ends of the years. These all coincide with the observable lunar crescents as tabulated for Babylon by Parker and Dubberstein,[84] except for a variation of one day in 322 and 321. The dates were not calculated for Athens, but are close enough for all practical purposes in this inquiry.[85]

337/6 (Phrynichos)
Ordinary 355: July 16—July 5

336/5 (Pythodelos)
Intercalary 384: July 6—July 24

φθίνοντος, καὶ ἄγοντες τὴν ἡμέραν ταύτην πάντα τὸν χρόνον, ἐσέβαλον ἐς τὴν 'Επιδαυρίαν καὶ ἐδῄουν. There were religious scruples against a campaign after the first day of the Doric sacred month of Karneios, which the Argives overcame by this device. This passage might be added to those given by Pritchett as examples of irregularity in B.C.H., LXXXI, 1957, pp. 276-282 (cf. Cl. Phil., XLII, 1947, p. 238).

[83] This happens to be a Metonic cycle in the astronomical calendar, though that fact is immaterial in our present study.

[84] Babylonian Chronology, pp. 35-36.

[85] I consider Pritchett's views as set forth in Cl. Phil., XLII, 1947, p. 243, too pessimistic. If the sequence of ordinary and intercalary years is known, and if the pattern of months and prytanies within the year is known, then it follows that the Julian date for any dated inscription is known (see also M. F. McGregor, A.J.P., LXXIX, 1958, p. 421 note 3). I make two assumptions: (1) that the normal date for Hekatombaion 1 was in July, and (2) that this date in every year after the fifth century was the same as Prytany I 1. I do not believe that the decision whether a year was to be ordinary or intercalary was irrevocable, though I agree with those who hold that such a decision always had to be made before the year began (cf. Pritchett, B.C.H., LXXXI, 1957, p. 298). The complex character of the year 307/6 shows, I think, a change within the year (Pritchett and Meritt, Chronology, p. 21), and in principle one must not hold the Athenians so rigidly bound as to deny them choice.

335/4 (Euainetos)
Ordinary 355: July 25–July 14

334/3 (Ktesikles)
Ordinary 354: July 15–July 2 (Leap Year)

333/2 (Nikokrates)
Intercalary 384: July 3–July 21

332/1 (Niketes)
Ordinary 354: July 22–July 10

331/0 (Aristophanes)
Ordinary 354: July 11–June 29

330/29 (Aristophon)
Intercalary 384: June 30–July 17 (Leap Year)

329/8 (Kephisophon)
Ordinary 354: July 18–July 6

328/7 (Euthykritos)
Intercalary 384: July 7–July 25

327/6 (Hegemon)
Ordinary 355: July 26–July 15

326/5 (Chremes)
Ordinary 354: July 16–July 3 (Leap Year)

325/4 (Antikles)
Intercalary 384: July 4–July 22

324/3 (Hegesias)
Ordinary 354: July 23–July 11

323/2 (Kephisodoros)
Ordinary 355: July 12–July 1 [86]

322/1 (Philokles)
Intercalary 384: July 2–July 19 [87] (Leap Year)

[86] See above, p. 132.
[87] *Ibid.*

321/0 (Archippos)

Ordinary 354: July 20–July 8

320/19 (Neaichmos)

Intercalary 384: July 9–July 27

319/8 (Apollodoros)

Ordinary 355: July 28–July 17

318/7 (Archippos)

Ordinary 354: July 18–July 5 (Leap Year)

The Aristotelian rule for lengths of prytanies holds in many instances, but is subject to variation, being at times completely reversed, and frequently showing the longer prytanies at the beginnings and ends of the years. But no rule is inviolable. One result of this freedom in distributing prytanies is that the dates by months in the festival calendar are more accurately predictable than the dates by prytany, except in those cases where there have been irregular intercalations.[88] Simple error in the record, whether in the festival dates or in the prytany dates, is of course unpredictable. The evidence of the fourth century is that there was no rigid pattern of prytanies. The attempt to substantiate such a pattern by citing rigid regularity of prytanies in the fifth century has proved vain. We must also examine the records of the third and second centuries to learn what sort of pattern, rigid or flexible, they too determine.

---

[88] This conclusion is the opposite of that reached by Pritchett and Neugebauer, *Calendars*, pp. 35-36: "in any case the prytany calendar advanced more regularly than the κατ' ἄρχοντα calendar and – – – when the κατὰ θεόν dates are lacking, as is generally the rule, the prytany calendar should be used to give the relative position of a day within the year." Their belief is based upon predicated irregularity in the festival calendar, which is rather a rarity than the general rule. But the validity of their thesis has a better chance near the dates of the important festivals, especially in Elaphebolion.

# CHAPTER VI

## THE TWELVE PHYLAI

Inasmuch as there were twelve prytanies from 307/6 down to 223, and again throughout the second century, it is obvious that they cannot have had 36 days each in the first part of the year and 35 days each in the second part of the year, no matter how the year may have been divided. But, following as closely as possible the scheme described by Aristotle for the fourth century, one might have expected, perhaps, that in the third and second centuries the normal festival year of 354 days had always six prytanies of 30 days followed by six prytanies of 29 days. There is, however, abundant epigraphical evidence that this scheme, though possible at times, was not followed in every ordinary year.

The literary tradition equated the twelve prytanies with the twelve months,[1] but if this was a general rule it was not universally followed. There are two preserved instances of a date κατὰ θεόν (so we know that the count had not been tampered) in the second century differing by a day from the prytany date and thereby proving that the month and prytany were not coterminous.[2] Moreover, prytanies of 30 days occur both early and late in the year, the date τριακοστή being known epigraphically from the 3rd, 9th, and 11th prytanies. It is also clear that

---

[1] Pollux, VIII, 115: πρυτανεία δέ ἐστι ὁ χρόνος ὃν ἑκάστη φυλὴ πρυτανεύει· καὶ ὅτε μὲν δέκα ἦσαν, πλείους ἑκάστη φυλῇ αἱ ἡμέραι, ἐπεὶ δὲ δώδεκα ἐγένοντο, ἑκάστη φυλὴ μηνὸς πρυτανείαν ἔχει. This passage is quoted by Pritchett and Neugebauer, *Calendars*, p. 78.

[2] One was in the year 190/89, to which I now assign (p. 235) the archon Demetrios and the text published by Pritchett and Meritt, *Chronology*, pp. 121-122. This text was first dated in 174/3 and later in 159/8; for its significance see Pritchett and Neugebauer, *Calendars*, p. 15 (no. 5) and p. 78. The other was in 164/3 (see below, p. 165 note 70).

135

in 304/3 the 12th prytany had 30 days.[3] And it can be demonstrated from calendar equations "that in 304/3 Prytany VI was
of 29 days, in 302/1 Prytany X of 29 days and Prytany XI of
30 days, and in 283/2[4] Prytany I of 30 days. On the other
hand, we have the conflicting evidence of years such as 306/5,
245/4, 195/4, and 169/8 where the prytany dates toward the
middle of the year were two or three days behind the civil dates,
as opposed to years such as 302/1, 250/49, and 187/6 where the
prytany dates were running ahead of those in the civil calendar.
In some of the former years, scholars have supposed that Prytanies I-VI were of 30 days length and Prytanies VII-XII of
29 days. In the latter examples, the first six prytanies have been
considered the 29-day ones." [5]

The evidence shows that there was in the third and second
century no one rigid rule for the arrangement of prytanies
within the conciliar year. The general statement of Pollux, who
formulated an apparently rigid rule, turns out to have the same
flexibility that was enjoyed by Aristotle's rule in the fourth
century. Some have felt "that the problem of the intercalation
and suppression of days in the civil calendar, as determined from
the fragmentary epigraphical evidence which usually furnishes
only two or three inscriptions for any one year, is too elusive
to permit the formulation of one rule for the sequence of 29-
and 30-day prytanies throughout these two centuries." [6]    On
the contrary, there is a rule, namely, one of flexible adherence
to or deviation from the norm of Pollux. Sometimes the festival
date fell behind the prytany date, sometimes it preceded it; we
must allow the inscriptions to speak for themselves.[7]

[3] Pritchett and Neugebauer, *Calendars*, p. 78.
[4] The archon Ourias, formerly assigned to 283/2, belongs to 281/0. See
*Hesperia*, XXIII, 1954, p. 314 (cf. p. 233, below).
[5] Pritchett and Neugebauer, *Calendars*, pp. 78-79.
[6] Pritchett and Neugebauer, *Calendars*, p. 79.
[7] Among the sources from which Pollux apparently drew for his constitu-
tional antiquities of Athens were the *Atthis* of Philochoros and, probably, the

In the former category there is an equation from the year 260/59 (archonship of Arrheneides) preserved by Diogenes Laertios,[8] in which the festival date is behind the prytany date by two days: Ἐπ' Ἀρρενείδου ἄρχοντος ἐπὶ τῆς Ἀκαμαντίδος πέμπτης πρυτανείας· Μαιμακτηριῶνος δεκάτῃ ὑστέρᾳ, τρίτῃ καὶ εἰκοστῇ τῆς πρυτανείας.

Maimakterion 21 = Prytany V 23 = 139th day.

This is a quite normal ordinary year, in which the months and prytanies were arranged as follows: [9]

Months      30 29 30 29 30 29 30 29 30 29 30 29 = 354
Prytanies   29 29 29 29 29 29 30 30 30 30 30 30 = 354.

In the year 247/6, when Diomedon was archon, there is an equation which shows the festival date to be again two days behind the prytany date: [10]

Ἐλαφηβολιῶνος ἕνει καὶ νέαι ἐμ[βολίμωι, δευτέραι τῆς] πρυτανείας.

Elaphebolion 30 = Prytany [X 2] = 266th day.

The year was an ordinary year, in which months and prytanies were arranged as follows:

Months      29 30 29 30 29 30 29 30 30 29 30 29 = 354
Prytanies   29 29 29 29 29 29 30 30 30 30 30 30 = 354.

---

Nomoi of Theophrastos (see A. E. Raubitschek, Classica et Mediaevalia, XIX, 1958, pp. 84-86). If Pollux had his description of the calendar of the twelve phylai from Theophrastos (Aristotle's successor), it is significant for our interpretation of Aristotle that the calendar of Theophrastos seems to give a fixed rule, but one which was in fact quite flexible.

[8] Lives, VII, 1, 10. For the date of Arrheneides, see below, pp. 221-226.
[9] See Pritchett and Neugebauer, Calendars, p. 81. The inscription I.G., II², 704, once assigned to this year, is now known to belong to the archonship of Lysitheides in 272/1 (cf. Dinsmoor, Hesperia, XXIII, 1954, p. 286 note 12, p. 314; Meritt, Hesperia, XXVI, 1957, p. 56).
[10] I.G., II², 791, lines 5-6.

An extra day was given to Elaphebolion, which had been planned as a hollow month, and from there to the end of the year the alternation of full and hollow months was reversed.

The year 187/6, when Theoxenos was archon, has an equation which shows the festival date to be once more two days behind the prytany date: [11]

Μουνιχιῶνος ἐνδεκάτηι, τ[ρίτηι καὶ δεκάτηι τῆ]ς πρυτανείας

Mounichion 11 = Prytany [X 13] = 277th day.

The year was an ordinary year, in which the months and prytanies were arranged as follows:

Months     30  29  30  29  30  29  30  29  30  29  30  29 = 354
Prytanies  29  29  29  29  29  29  30  30  30  30  30  30 = 354.

These examples illustrate one arrangement, which I hold quite normal (one of many), both of months and prytanies, within an ordinary year. But it is, of course, possible to avoid the straightforward interpretation of these texts, if one wants to assume tampering with the festival calendar. The *advocatus diaboli* could suggest in the last instance, for example, that two days were added irregularly somewhere before Mounichion 11 and removed somewhere thereafter, and that the normal months (κατὰ θεόν) and prytanies should all be restored according to the month-by-month, *pari passu*, formula of Pollux.

It is more difficult to avoid the implications of an equation where the festival date is in advance of the prytany date. For example, in the festival year 306/5, the first six prytanies had 30 days each, and the last six prytanies would have had 29 days each except that an extra day was given to Mounichion (*I.G.*, II², 471) thus making a year of 355 days, and adding a day to one of the last three prytanies, making it have 30 instead of 29

[11] Pritchett and Meritt, *Chronology*, p. 117.

days. This year has been the subject of a special study by W. K. Pritchett,[12] whose reconstruction seems to me impeccable, though I should now read the first four months as 29, 30, 29, 30 in order to have hollow Hekatombaion follow full Thargelion and Skirophorion of 307/6 (see below, p. 177).

Also, in the year 195/4, a decree passed on the 27th day of some prytany praising the epheboi of the preceding year, and hence datable to the third prytany and to the month Boedromion, yields the equation:[13]

$$[\text{Βοηδρομιῶνος } ἔνει \; καὶ \; νέαι], \; ἑβδόμει \; καὶ \; εἰ[κοστεῖ \; τῆς \\ πρυτανείας]$$

$$[\text{Boedromion } 29/30] = \text{Prytany [III] } 27 = 88\text{th day.}$$

Even if one were to assume that the first three months of the year had only 29 days each, the first two prytanies will have had each 30 days, and there is no possibility that one "might postulate prytanies alternating between 30 and 29 days."[14] Nor can the assumption of three months of 29 days each in sequence be accepted. The idea that there might be two (or even three) hollow months together depends on the hypothesis that months in Athens were determined by actual observation, each month, of the new lunar crescent, a hypothesis which we found (above, pp. 16-37, 44) not valid for Athens. Moreover, the visibility of the new crescent marked the first three months in the year 195/4, in Babylon at least, as 30 29 30, and not as 29 29 29. Parker and Dubberstein give the Julian dates, to which I have added the names and lengths of the Athenian months:[15]

[12] *A.J.P.*, LVIII, 1937, pp. 329-333.
[13] Pritchett and Meritt, *Chronology*, p. 111.
[14] Cf. Pritchett and Neugebauer, *Calendars*, p. 78. One should also note that Prytany IV (an even number) is known to have had a 30th day in 178/7 (*Hesperia*, Suppl. I, no. 64, line 2), unless the text of the inscription is in error. See Pritchett and Neugebauer, *Calendars*, pp. 87-88, where in the first equation Μαιμακτηριῶνος ἕκτει ἰσταμένου, δεκάτει τῆς πρ[υ]τ[α]νείας is an error for Πυανο-ψιῶνος ἐνάτει μετ' εἰκάδας, τριακοστεῖ τῆς πρυ[τανε]ίας.
[15] *Babylonian Chronology*, p. 40.

| 195 B.C. | June 8 | } | Skirophorion | 29 |
| | July 7 | | | |
| | | } | Hekatombaion | 30 |
| | August 6 | | | |
| | | } | Metageitnion | 29 |
| | September 4 | | | |
| | October 4 | } | Boedromion | 30 |

The vagaries of the moon cannot have satisfied these figures for Babylon and then have produced a set of months shorter over all by two days in Athens. One refuge, of course, is to say that the observations in Athens were faulty. If faulty, the observation for Hekatombaion 1 must be taken, at least, as correct, for here the prytany year and the festival year were coterminous, and the date of Hekatombaion 1 κατ᾽ ἄρχοντα never differs from the date of Hekatombaion 1 κατὰ θεόν.[16] This means, of course, that the faulty observations were for Metageitnion 1 and Boedromion 1, the new crescent moon for the end of the second month being "observed" two days before it was, in fact, astronomically observable.

These are desperate measures, and are necessary only to save a hypothetically rigid scheme that even at best does not follow any one pattern of prytany sequence. In the present instance, the evidence is that one of the first two prytanies, at least, in this ordinary year had 31 days. It was so understood by Pritchett and me when we published our *Chronology* in 1941, and I so understand it still.

A similar type of calendar equation would seem to occur in the archonship of Thymochares, and this is true if his year was

[16] This is implicit also in the hypothesis of Pritchett and Neugebauer, *Calendars*, p. 18: "Nevertheless, a calendar κατὰ θεόν was still necessary, not only to keep the calendar in some sort of relation with the moon, but also to establish terminal limits for the civil years after the passage of 12 (or 13) real months, regardless of the right of archons or demos to tamper with the calendar within any particular year." But see below, pp. 172-175.

ordinary. Recent adjustments in the calendar of the third century make it extremely probable that Thymochares belongs in 258/7, a date for which an ordinary year is eminently suitable.[17] The equation from *I.G.*, II², 700, has been read as [Βοηδ]ρομιῶνος ἕνει καὶ [νέαι, ἑβδόμει καὶ εἰκοστεῖ τῆς πρυτανείας] :

Boedromion 29 = Prytany [III 27] = 88th day.[18]

The equation is restored, but the text is stoichedon, and the only two correct restorations, epigraphically, give the date within the prytany, as above, for an ordinary year, or as πέμπτει καὶ εἰκοστεῖ for an intercalary year. The suggestion has been made [19] that we cannot insist on perfect stoichedon order but that one "may equally well restore [ᵛ ἐνάτει καὶ εἰκοστεῖ], for an ordinary year." Kirchner's note on *I.G.*, II², 700, is cited as authority that stoichedon order was frequently violated. What Kirchner says is "V. non omnes στοιχηδόν exarati neque eiusdem longitudinis sunt." Stoichedon order was violated in line 10 and in line 18, in each of which an extra letter more than the normal 50 was crowded in. This is not "frequent" violation, nor does it argue that one should restore the date with one letter less (rather than more) than normal. The uninscribed spaces were not used at random in this text, but purposefully. They separate major clauses, or names and titles. There is, in fact, no example of an uninscribed space in the preamble, and one can only assume one (if he wishes to do so) by way of an exception.

If the year was ordinary the equation given above must stand, and one of the first two prytanies had 31 days. Pritchett and Neugebauer would claim two prytanies of 30 days and three consecutive hollow months, but the objections to three consecu-

---

[17] See Dinsmoor, *Hesperia*, XXIII, 1954, p. 314; Meritt, *Hesperia*, XXVI, 1957, p. 97.

[18] See *Hesperia*, VII, 1938, pp. 110-112; Pritchett and Meritt, *Chronology*, p. 97.

[19] Pritchett and Neugebauer, *Calendars*, p. 81 note 9.

tive months of 29 days, as registered above for 195/4, are valid also in 258/7. However, one could restore an intercalary year with the equation [Βοηδ]ρομιῶνος ἔνει καὶ [νέαι, πέμπτει καὶ εἰκοστεῖ τῆς πρυτανείας].

Boedromion 30 = Prytany [III 25] = 89th day.

In the archonship of Theophemos (245/4)[20] the festival calendar was also running ahead of the prytany calendar. The text of *I.G.*, II², 795, is as follows:[21]

*a.* 245/4 *a.* ΣΤΟΙΧ. 41

Ἐπὶ Θεοφήμου ἄρχοντος ἐπὶ [τῆς ....⁸.... τετάρτης]
πρυτανείας ἧι Προκ[λ]ῆς Ἀπ[.........¹⁵....... ἔγραμ]
[μά]τευεν· Πυανοψιῶνος ἔκ[τει ἐπὶ δέκα, τετάρτει καὶ]
[δεκ]άτει τῆς πρυτανε[ίας· – – – – κτλ. – – – – – ]

Pyanepsion [1]6 = Prytany [IV] 1[4] = 104th day.

This is suitable for an ordinary year in which the first three prytanies had 30 days each, and it agrees with the evidence, as restored, of *I.G.*, II², 799, from the same year:[22]

*a.* 245/4 *a.* ΣΤΟΙΧ. 52

[θ] ε ο ί
['Επὶ Θεοφήμου ἄρχοντος ἐπὶ τ]ῆς Ἐρεχ[θ]εῖδος τρίτης πρυτανείας ἧ
[ι Προκλῆς Ἀπ......¹⁵...... ἐ]γρ[α]μμάτευεν· ᵛ Βοηδρομιῶνος ἔν
[ηι καὶ νέαι, ὀγδόηι καὶ εἰκοστῆι τῆ]ς πρυτα[ν]είας· ᵛ – – κτλ. – –

Boedromion [29] = Prytany III [28] = 88th day.

The probable arrangement of months and prytanies within the year was as follows:

[20] For the date see Pritchett and Meritt, *Chronology*, p. xxii; Meritt, *Hesperia*, XVII, 1948, p. 13; W. B. Dinsmoor, *Hesperia*, XXIII, 1954, p. 315.
[21] See Meritt, *Hesperia*, IV, 1935, p. 551.
[22] See Meritt, *Hesperia*, IV, 1935, p. 550.

Months   29 30 29 30 29 30 29 30 29 30 29 30 = 354
Prytanies 30 30 30 30 30 30 29 29 29 29 29 29 = 354.

But only the first three months and the first three prytanies are really attested.[23]

Possibly a similar arrangement of prytanies was valid for the archonship of Theophilos (227/6) where the text of *I.G.*, II², 837, should, I think, be restored as follows:

*a*. 227/6 *a*.                                       NON-ΣTOIX. 40-42

[θ]                  ε               ο                        [ι]

['Επὶ] Θεοφίλου ἄρχοντος ἐπὶ τῆς Κεκροπίδ[ος τρίτης]

[πρυτ]ανείας ἧι Φίλιππος Κηφισοδώρου 'Αφ[ιδναῖος]

[ἐγρα]μμάτευεν· Βοηδρομιῶνος ἕκτει μετ᾽ [εἰκάδας, τε]

5   [τάρτ]ει καὶ εἰκοστεῖ τῆς πρυτανείας· – – [ – κτλ. –]

Boedromion 25 = Prytany [III] 2[4] = 84th day.

The lines are syllabically divided, and so not of equal length. Here the first two prytanies were of 30 days each, and Hekatombaion was full, Metageitnion hollow, and Boedromion full. This is not a necessary, but it is a normal (and probable), arrangement.

In the archonship of Eunikos (169/8) the text of *I.G.*, II², 910,[24] shows that the first six prytanies were of 30 days each:

*a*. 169/8 *a*.                                       NON-ΣTOIX.

['E]πὶ Εὐνίκου ἄρχοντος ἐπὶ τῆς Οἰνεῖδος^{vv[vvv]} ἐ[β]δόμης π[ρυτα]
νείας ἧι Ἱερώνυμος Βοήθου Κηφισιεὺς ἐγραμμάτευεν· Γαμηλι[ῶνος]
[ἕ]κτει μετ᾽ εἰκάδας, δευτέραι [καὶ εἰκοστεῖ] τῆ[ς πρυτανείας ^{vv}]
κτλ.

Gamelion 25 = Prytany VII [2]2 = 202nd day.[25]

[23] For the calendar of the following year (Kydenor, 244/3), see below, pp. 146-148.

[24] For the text see Dow, *Hesperia*, Suppl. I, pp. 129-133 (71).

[25] See Meritt, *Hesperia*, IV, 1935, p. 558. I would now interpret the year as ordinary, beginning with full Hekatombaion.

Here the months and prytanies were arranged as follows:

Months     30  29  30  29  30  29  30  29  30  29  30  29 = 354
Prytanies  30  30  30  30  30  30  29  29  29  29  29  29 = 354.

In *Hesperia*, V, 1936, p. 429 (17), the last day of Skirophorion in this year was equated with the 29th day of the twelfth prytany.

These texts, and others, from the second and third centuries, show ordinary years, which must be considered normal, in which the prytanies were so arranged that all the 30-day prytanies came together at the beginning of the year. We now have the complement to that order of prytanies discussed above which was just the reverse. And both series are in addition to those years wherein the prytanies followed the months (approximately), alternating between 29 and 30 days. Once surely, and probably twice, there has been evidence for a prytany of 31 days. If there is validity in assuming a kind of continuity of practice down through the fifth, fourth, third, and second centuries, the variety of types of conciliar year (with lack of rigidity) in the fifth, third, and second centuries argues for variety also in the fourth century, where Aristotle's "rule" $(4 \times 36 + 6 \times 35)$ may have as many modifications as the evidence warrants, and confirms the definition of flexibility in the fourth century as set forth in the preceding chapter. Even a prytany of 37 days, I think, might be accepted there as an unusual but possible variation.

I have suggested that the year 176/5, being ordinary, may have had one prytany of 31 days.[26] Such apparently was the case in 195/4 and probably also in 258/7.[27] Pritchett has re-

---

[26] *Hesperia*, XXVI, 1957, p. 69.
[27] See above, pp. 139-140, 141. Pritchett and Neugebauer, *Calendars*, p. 76, state it as dogma that "a 31-day prytany is not possible in an ordinary year of the period of twelve phylae."

affirmed his preference for rigidity in the prytany calendar,[28] and in particular he chooses a very large distortion in the festival year (for which he can cite parallels) to a minor variation of one day in the prytany calendar (which he does not think possible). It is the possibility of such a variation in the prytany calendar that we have been considering here.

I should also claim, in principle, that there may have been at times a prytany of 31 or of 33 days,[29] instead of the consistently regular 32 which were normal, in an intercalary year. But I have as yet found no formal evidence for such a variation in the preserved documents. A close approach to a demonstration of it seemed to lie in the comparative study of *I.G.*, II², 780A and 777, two texts from the archonship of Kallimedes now attributed to the year 252/1.[30] The readings in the *Corpus* are as follows:

*I.G.*, II², 780A

*a*. 252/1 *a*.                                     NON-ΣΤΟΙΧ.

M     ο     υ̂     σ     [α     ι]

Ἐπὶ Καλλιμήδου ἄρχοντος ἐπὶ τῆς Αἰαντίδος ἐνάτης πρυτανεία[ς ἧι Καλ]

λίας Καλλιάδου Πλωθεὺς ἐγραμμάτευεν· Ἐλαφηβολιῶνος δεκάτηι

[ὑστέρα]

ι, ἐνάτηι καὶ εἰκοστῆι τῆς πρυτανείας· – – – – – κτλ. – – – – – – –

Elaphebolion 2[1] = Prytany IX 29 [31]

*I.G.*, II², 777

*a*. 252/1 *a*.                                      ΣΤΟΙΧ. 42

Ἐπὶ Καλλιμήδου [ἄρχοντος ἐπὶ τῆς . . . . .¹¹ . . . . . . δεκά]

[τη]ς πρυτανείας ἧ[ι Καλλίας Καλλιάδου Πλωθεὺς ἐγρα]

---

[28] *B.C.H.*, LXXXI, 1957, pp. 272-273.

[29] Quite apart from those adjustments at the end of the year which Pritchett and Neugebauer think possible (*Calendars*, p. 68).

[30] Pritchett and Meritt, *Chronology*, p. xxi. See below, p. 234.

[31] The texts are accepted and the calendar equations so given also by Pritchett and Neugebauer, *Calendars*, p. 70.

[μμάτ]ενεν· Μουνιχι[ῶνος δωδεκάτηι, ὀγδόηι καὶ δεκάτ]
[ηι τῆς] πρυτανεία[ς· — — — — — — κτλ. — — — — — —]
Mounichion [12] = Prytany [X 18] [32]

If these equations are valid, then Kirchner's judgment (commentary on *I.G.*, II², 777) is sound: "Annus est intercalaris ab Hecatombaeone cavo incipiens, quo prytania I = 33, II-XI 32, XII 31 dies." But the restoration at the end of line 3 and beginning of line 4 in *I.G.*, II², 780A, should be Ἐλαφηβολιῶνος δεκάτηι [προτέρα]ι, ἐνάτηι καὶ εἰκοστῆι τῆς πρυτανείας.[33] The text is not written stoichedon, but this is preferable, I think, so far as the available space is concerned, to the restoration [ὑστέρα]ι, and it enables one to make the equation:

Elaphebolion 2[0] = Prytany IX 29 = 285th day.

The text of *I.G.*, II², 777, must now be restored either as Μουνιχι[ῶνος ἐνδεκάτηι, ὀγδόηι καὶ δεκάτηι τῆς] πρυτανεία[ς] or as Μουνιχι[ῶνος δωδεκάτηι, ἐνάτηι καὶ δεκάτηι τῆς] πρυτανεία[ς]. The possible equations are:

Mounichion [11] = Prytany [X 18] = 306th day
Mounichion [12] = Prytany [X 19] = 307th day.

Months and prytanies throughout the year are to be arranged, with no irregularities, as follows:

Months      29  30  29  30  29  30  29  30  29  30  29  30  30 = 384
Prytanies   32  32  32  32  32  32  32  32  32  32  32  32 = 384.

Again, in the archonship of Kydenor (244/3),[34] there is now a firm equation which was not available when Pritchett and Neugebauer made their study, and which seems to demand at

[32] Pritchett and Neugebauer, *Calendars*, p. 70.
[33] For the phrase δεκάτη προτέρα to mean the 20th day, see above, pp. 46, 58.
[34] For the date see Meritt, *Hesperia*, XVII, 1948, p. 13; below, p. 234.

least one prytany of 33 days in an intercalary year. The opening lines of the new decree are as follows: [35]

Hesperia, XVII, 1948, pp. 3-4 (3)

a. 244/3 a.                                          NON-ΣΤΟΙΧ.

θ            ε            ο            ί

Ἐπὶ Κυδήνορος ἄρχοντος ἐπὶ τῆς Ἐρεχθεῖδος ἐνάτης πρυ
τανείας ἧι Πολυκτήμων Εὐκτιμένου Εὐπυρίδης ἔγραμ
μάτευεν· ᵛ Ἐλαφηβολιῶνος ἐνάτηι ἱσταμένου, ἑβδόμηι καὶ
5  δεκάτηι τῆς πρυτανείας· ἐκκλ[η]σία κ[υ]ρία· – κτλ. – –

Elaphebolion 9 = Prytany IX 17 = 251st day.

This year has, until now, been interpreted as intercalary; but if the prytanies had regularly 32 days each the year can have advanced only so far as the 273rd day by the time of this decree. In order to equate this day with Elaphebolion 9 one must assume that somewhere in the festival calendar there were successively at least three hollow months of 29 days each, or, barring this, two hollow months in succession twice, before Elaphebolion. There is no evidence that such a sequence ever existed at Athens, at least not in historical times. Normally, the festival calendar would bring Elaphebolion 9 either to the 274th or to the 275th day of the year. In an intercalary year the alternative is to assume that one, at least, of the first eight prytanies had 33 days. If the year had to be intercalary I should now make this assumption and cite the year as having such a prytany of irregular length. Since I hold no brief against this in principle, and indeed think that irregular lengths are at times demonstrable in ordinary years, it would seem a satisfactory solution, and I should maintain it here except for the fact that the festival date is Elaphebolion 9, just before the City Dionysia, and at that time of year

---

[35] Hesperia, XVII, 1948, pp. 3-4 (3). The decree published by Pritchett and Meritt, Chronology, pp. 23-27, was passed on the same day.

when irregular additions to days of the festival calendar are most apt to occur.[36] If one interprets the year as ordinary, a normal progression of months and prytanies is possible except for a dislocation and its correction in Elaphebolion:

Months     29 30 29 30 29 30 29 30 29 30 29 30 = 354
Prytanies  29 29 29 29 29 29 30 30 30 30 30 30 = 354.

In this case the 17th day of the ninth prytany was the 251st day of the year, whereas the festival calendar with a normal Elaphebolion 9 would have advanced only so far as the 245th day. This implies that six days had been intercalated in the festival calendar, possibly early in Elaphebolion, and possibly in order to delay the celebration of the City Dionysia. I assume that the compensation by omission of days was made after the festival but before the end of the month. This seems to me preferable to the assumption of an irregular length of prytany in an intercalary year, or to the supposition that there were three hollow months in succession in the festival calendar.[37] The other known text of Kydenor's year which contains a calendar equation should now be restored as of an ordinary year: I suggest the following as a possible, though not the only possible, solution: [38]

*I.G.*, II², 766 + *Hesperia*, XVII, 1948, pp. 5-7

a. 244/3 a.                                NON-ΣTOIX.

[Ἐπὶ Κυδήνορος ἄρχοντος ἐ]πὶ τῆς ▨[▨▨▨▨▨ δος ἕκτης πρυτα]
[νείας ἧι Πολυκτήμων Εὐκτι]μένου Εὐ[πυρίδης ἐγραμμάτευεν· Πο]
[σιδεῶνος δωδεκάτηι, τετάρ]τηι καὶ δ[εκάτηι τῆς πρυτανείας· ἐκ]
[κλησία κυρία· τῶν προέδρων ἐ]πεψήφιζε[ν — — — — κτλ. — — — —]

[Posideon 12] = Prytany [VI] 1[4] = 159th day.

[36] See below, pp. 161-165.
[37] This is not the solution proposed in *Hesperia*, XVII, 1948, p. 4.
[38] See *Hesperia*, XVII, 1948, p. 5, for the restoration as of an intercalary year.

Still another text must be considered in our search for an irregular length of prytany, if evidence of such has been preserved, in an intercalary year. I have recently drawn the conclusion from three equations, close together in time, in the archonship of Euthykritos (189/8), that the eighth prytany of the year had only 31 days.[39] The equations are:

Prytany [VIII] 31 = Anthesterion 19
Prytany IX 8 = Anthesterion 27
Prytany IX 21 = [Elaphebolion 10]

This view has been challenged,[40] but in spite of this I still hold an irregular length of prytany in an intercalary year to be theoretically possible and worth consideration. If there was actually an intercalation of an extra day in the festival calendar of 189/8 between Anthesterion 19 and 27,[41] Anthesterion 27 falling on Prytany IX 8, then Prytany IX 21, had the compensating omission been made early in Elaphebolion, would have fallen on Elaphebolion 11. This is difficult, because it does not square with the epigraphical requirements of the texts.[42] There is the same difficulty if the second date in Anthesterion ($\tau\epsilon\tau\rho\grave{a}s$ $\mu\epsilon\tau$' $\epsilon\grave{i}\kappa\acute{a}\delta as$) is taken as the 26th day of a hollow month. Now, not one but two days have to be assumed as intercalated between the 19th and the 26th so that with regular prytanies of 32 days each Anthesterion 26 may fall on Prytany IX 8. Supposing the compensation to have been made in early Elaphebolion, we find that Prytany IX 21 would have fallen on Elaphebolion 12. This is no better than Elaphebolion 11 for the epigraphical requirements. If one assumes that the compensation was made only after Prytany IX 21, there is the difficulty of carrying the

[39] Hesperia, XXVI, 1957, p. 65.
[40] W. K. Pritchett, B.C.H., LXXXI, 1957, p. 272.
[41] As demanded by those who wish to keep regular lengths of prytanies.
[42] Hesperia, XXVI, 1957, pp. 63-64 (17, lines 25-26). See also especially p. 65 note 22.

irregularity over a rather long span of days, but this is not a formal barrier to the assumption. Insisting on a prytany of 32 days, one should probably take Anthesterion as hollow and assume the intercalation of two days with compensation after the Dionysiac festival by omitting two days from full Elaphebolion. Months and prytanies were arranged, possibly, as follows:

$$\text{Months} \quad 29 \ \ 30 \ \ 29 \ \ 30 \ \ 29 \ \ 30 \ \ 29 \ \ 30 \ \ \overset{+2}{29} \ \ \overset{-2}{30} \ \ 30 \ \ 29 \ \ 30 = 384$$
$$\text{Prytanies} \quad 32 \ \ 32 \ \ 32 \ \ 32 \ \ 32 \ \ 32 \ \ 32 \ \ 32 \ \ 32 \ \ 32 \ \ 32 \ \ 32 = 384.$$

But I regard the problem of this year as by no means surely solved.

In summary, it may be said that we have found ample evidence for differing sequences of prytanies in ordinary years, with occasional irregularities of a prytany longer by one day than the normal higher limit of 30 days, but that (however acceptable in principle) we have not found formal proof that any prytany in an intercalary year ever varied from the normal length of 32 days.

The remainder of this chapter is given over to individual problems of restoration, reading, and interpretation, based upon texts that show some calendrical irregularity or other noteworthy peculiarity.

I

Since the publication of *The Calendars of Athens* a new text has been added to those already known from the year 303/2,[43] confirming a normal calendar count in an intercalary year:

*Hesperia*, XXI, 1952, pp. 367-368 (8)

     *a*. 303/2 *a*.                  ΣΤΟΙΧ. 29

     [Ἐπὶ Λεωστράτου ἄρ]χοντ[ος ἐπὶ τῆς Αἰ]

---

[43] See Pritchett and Neugebauer, *Calendars*, p. 69.

[αντίδος δωδεκάτ]ης πρυτ[ανείας ἧι Δ]
[ιόφαντος Διονυσ]οδώρου Φ[ηγούσιος]
[ἐγραμμάτευεν· Σκ]ιροφορ[ιῶνος ἔνει]
5  [καὶ νέαι, δευτέρα]ι καὶ τρ[ιακοστεῖ τ]
[ῆς πρυτανείας· — ] — — κτλ. — [ — — ]

II

The calendar character of the two years 295/4 and 294/3 has been studied above (pp. 26-33), with observations on the intercalation of extra days early in Elaphebolion in order to delay the celebration of the City Dionysia.

III

The classic example of retardation of the festival calendar by the addition of extra days is now the decree from the archonship of Pytharatos (271/0) published by W. B. Dinsmoor in *Hesperia*, XXIII, 1954, pp. 299-300:

Ἐπὶ Πυθαράτου ἄρχοντος ἐπὶ τῆς Λεωντίδο[ς]
[ἐ]νάτης πρυτανείας ἧι Ἰσήγορος Ἰσοκράτο[υ]
[Κ]εφαλῆθεν ἐγραμμάτευεν· Ἐλαφηβολιῶνο[ς]
[ἐ]νάτει ἱσταμένου τετάρτει ἐμβολίμωι, ἐβδ[ό]
5  [μ]ει καὶ εἰκ[οσ]τεῖ τῆς πρυτ[α]νείας· ἐκκλησί[α·]
κτλ.

Elaphebolion 9 (intercalated the fourth time) = Prytany IX 27

Another inscription published by Dinsmoor at the same time shows that the year of the archon Pytharatos was intercalary:

*Hesperia*, XXIII, 1954, pp. 288-289

a. 271/0 a.                    NON-ΣTOIX.

[θ]          ε          o          ί
Ἐπὶ Πυθαράτου ἄρχοντος ἐπὶ τῆς Ἀντιγονί
δος δευτέρας πρυτανείας ἧι Ἰσήγορος Ἰσοκρά

τοῦ Κεφαλῆθεν ἐγραμμάτευεν· Μεταγειτνιῶ
5 νος ἐνάτει ἱσταμένου, ἑβδόμει τῆς πρυτανεί
ας· — — — — — — — κτλ. — — — — — — — — — — —

Metageitnion 9 = Prytany II 7 = 39th day.

The equation is normal for an intercalary year in which the
prytanies had 32 days each and the first month had 30 days.
But in the month of Elaphebolion the ninth day, normally,
would have been only the 275th day of the year, whereas the
27th day of the ninth prytany should have been the 283rd day.
The discrepancy was caused in part by delaying the progression
of the festival year with four successive intercalations after
Elaphebolion 9, the fourth of them (without further help)
reaching as far as the 279th day. In order to achieve the equation
of the epigraphical text, therefore, it must be assumed that there
were other days (four in number) intercalated before Elaphebolion 9. These extra days (eight in all) may indeed all have
been added early in Elaphebolion and deleted within the same
month after the Dionysiac festival.[44] Apparently the purpose
of the intercalations was to delay the celebration of the festival.

IV

The year 226/5 is instructive not only as offering a clear
example of backward count with the phrase μετ᾽ εἰκάδας,[45] but
also as showing the difficulty of fixing the time of intercalations
of extra days and of the compensations for them. The two
calendar equations of this year (I do not repeat the Greek texts)
are:

Metageitnion 22 (intercalated the second time) = Prytany III 20
Metageitnion 29                                      = Prytany III 27

---

[44] See below, pp. 192-195.
[45] See B. D. Meritt, *Hesperia*, IV, 1935, p. 531. My interpretation of δευτέραι
ἐμβολίμωι was incorrect, but the proof of backward count was sound.

It follows that Metageitnion 22, intercalated the second time, must have been actually the 24th day of the month (if there was no other irregularity) and that Metageitnion 29 must have been actually the 31st day (seven days later).[46] Pritchett and Neugebauer have used the normal progression of twelve prytanies of 32 days each in an intercalary year (in this case the extra month was a second Hekatombaion) as a frame against which to set the months of the festival calendar, which they arrange as follows:

30   30   30+2   29   29   29   29   30   etc. – – – – .

This scheme protracts the irregularity unduly. And we prefer not to have three months planned with 30 days each at the beginning of the year. I suggest instead the sequence:

30   29   30+3   29−3   30   29   30   29   etc. – – – – ,

where the irregularity, introduced in the latter part of the third month, was corrected in the early part of the fourth month. If, for example, Μεταγειτνιῶνος δεκάτη ὑστέρα was followed by δεκάτη ὑστέρα ἐμβόλιμος, the 21st day (intercalated) would actually have been the 22nd day, the 22nd day (so named) would actually have been the 23rd day, the 22nd day (intercalated) actually the 24th, and the 22nd day (intercalated a second time) actually the 25th. After seven more days had elapsed the 29th day (so named) would have been actually the 32nd. But the irregularity and its correction can be brought down within the compass of about two weeks and not extended over a span of six months, indeed, longer than six months, for there were still one or two days to be subtracted from the festival year after the 8th of Elaphebolion, which Pritchett and Neugebauer equate with Prytany VIII 22 and fix as the 246th day of the year.[47]

[46] See Pritchett and Neugebauer, *Calendars*, pp. 73-74.
[47] Pritchett and Neugebauer, *Calendars*, p. 70.

The evidence for this last equation is in an inscription from the Agora at Athens first published by Pritchett in *A.J.P.*, LXIII, 1942, p. 422:

*a*. 226/5 *a*.                                    NON-ΣTOIX. *ca.* 50

['E]πὶ 'Εργοχάρου ἄρχοντ[ος ἐπὶ τῆς Αἰαντίδος ὀγδόης πρυτανείας]
ἧι Ζωΐλος Διφίλου 'Αλωπ[εκῆθεν ἐγραμμάτευεν· 'Ανθεστηριῶνος]
ὀγδόει ἱσταμένου, δευτ[έραι καὶ εἰκοστεῖ τῆς πρυτανείας· ἐκκλησί]
α ἐν τῶι θεάτρωι· – – [– – – – – – – – κτλ. – – – – – – – – –]

But, as Pritchett noted,[48] other months than Anthesterion and other prytanies than the eighth may be restored. If the month was Boedromion and the prytany the fourth, then the equation is possible with these restorations:

[Boedromion] 8 = Prytany [IV] 2

*a*. 226/5 *a*.                                    NON-ΣTOIX. *ca.* 50

['E]πὶ 'Εργοχάρου ἄρχοντ[ος ἐπὶ τῆς – – – – – τετάρτης πρυτανείας]
ἧι Ζωΐλος Διφίλου 'Αλωπ[εκῆθεν ἐγραμμάτευεν· Βοηδρομιῶνος]
ὀγδόει ἱσταμένου, δευτ[έραι τῆς πρυτανείας· ἐκκλησία κυρί]
α ἐν τῶι θεάτρωι· – – – [– – – – – – – – κτλ. – – – – – – –]

Two of the days of correction were omitted in Boedromion before the eighth and one after the eighth, and the month which should have had 29 days had only 26, thus correcting the discrepancy by which the preceding month, which should have had only 30 days, had 33.

v

Among the ordinary years of the second century, it has been assumed recently that a very considerable dislocation in the festival calendar occurred in 188/7.[49] The evidence lay in the

---

[48] *A.J.P.*, LXIII, 1942, p. 422 note 34.
[49] Pritchett and Neugebauer, *Calendars*, pp. 29-30, 84.

text of *I.G.*, II², 891, in which the 11th or 12th of some month was equated with the 18th day of a prytany. Though neither month-name nor prytany-name is preserved, the equation was held to be evidence that the festival count had been retarded by 6 or 7 days. But the text of the *Corpus* was here in error and has since been corrected.[50] The inscription gives a quite different equation, with a text which I restore as follows:

*I.G.*, II², 891

a. 188/7 a.                                                     NON-ΣTOIX.

['Επὶ Συμμάχου ἄρχοντος ἐπὶ τῆς – – – – δευτέρας πρυ]τανείας ἧι
    'Αρχικλῆς [Θε]

[οδώρου Θορίκιος ἐγραμμάτευεν· Μεταγειτνιῶνος ἐ]νάτει ⟨ἐπὶ δέκα⟩,
    ὀγδόει καὶ δεκάτει τῆς πρυ

[τανείας· ἐκκλησία κυρία ἐν τῶι θεάτρωι· – – – – – – κτλ. – – – – – –].[51]

Metageitnion ⟨1⟩9 = Prytany [II] 18 = 48th day.

In this year the question also arises of forward count with μετ᾿ εἰκάδας, and normal sequences of months and prytanies may be restored if it is adopted.[52] The next equation depends too much on restoration to be significant, and is possible either with forward or with backward count. I have elected to show forward count in my restoration because of the need for forward count later in Mounichion (*I.G.*, II², 892; see below). The text now in question reads:

*I.G.*, II², 890

a. 188/7 a.                                                     NON-ΣTOIX.

['Επὶ Συμμά]χου ἄρχοντος ἐπὶ [τῆς – – – – – – ἕκτης πρυτανεί]
[α]ς ἧι 'Αρ[χικλ]ῆς Θεοδώρου Θορίκ[ιος ἐγραμμάτευεν· δήμου ψή]

---

[50] *A.J.P.*, LXXVIII, 1957, p. 381.
[51] For the date within the month see *A.J.P.*, LXXVIII, 1957, p. 381.
[52] See above, pp. 53, 58-59.

156 The Twelve Phylai

[φισ]μα· Ποσι[δεω]νος [ἔ]κ[τ]ει μετ᾽ εἰκάδα[ς, ὀγδόει καὶ εἰκοστεῖ
τῆς πρυ]
τ[αν]είας· ἐ[κκλη]σία ἐμ Πειραιεῖ————[—————κτλ.—————]⁵³

Posideon 26 = Prytany [VI 28] = 174th day.

Months and prytanies within the year may have been arranged
as follows:

Months     29 30 30 29 30 29 30 29 30 29 30 29 = 354
Prytanies  30 29 29 29 29 29 29 30 30 30 30 30 = 354.

The next equation shows that the festival calendar had been
retarded by three days, a divergence from the prytany calendar
which is normal if there was an early succession of 29-day
prytanies:

Hesperia, XV, 1946, p. 145 (6)

a. 188/7 a.                                           NON-ΣΤΟΙΧ.

Ἐπὶ Συμμάχου ἄρχοντος ἐπ[ὶ τῆς —————ὀγδόης πρυτανείας ἧι ᾽Αρχι]
[κ]λῆς Θεοδώρου Θορίκιος ἐ[γραμμάτευεν· δήμου ψήφισμα· ᾽Ανθε-
στηριῶνος]
ὀγ[δ]όει ἐπὶ δέκα, μιᾶι κα[ὶ εἰκοστεῖ τῆς πρυτανείας· ἐκκλησία κυρία
ἐν τῶι
θε[ά]τρωι· —————— [——————————κτλ. ——————————]

[Anthesterion] 18 = Prytany [VIII] [2]1 = 225th day.

The next equation belongs in Mounichion, and necessitates
forward count with μετ᾽ εἰκάδας if there is to be no irregularity
in either festival or conciliar year:

---

⁵³ The text is known from Fourmont's copy; cf. *C.I.G.*, 112. Sterling Dow,
*Hesperia*, Suppl. I, p. 108, suggested the division between lines 3 and 4 as
[πρυτα]|νείας. I note that there should also be syllabic division between lines
1 and 2.

*I.G.*, II², 892

*a.* 188/7 *a.*                    NON-ΣΤΟΙΧ.

['Ε]πὶ Συμμάχου ἄρχον[τος ἐπὶ τῆς — — — —]
δος δεκάτης πρυτα[νείας ἧι 'Αρχικλῆς]
Θεοδώρου Θορίκιος ἐγρ[αμμάτευεν· Μου]
νιχιῶ[νο]ς δευτέραι μετ' [εἰκάδας, τετάρτει]
5 καὶ εἰ[κ]οστεῖ τῆς πρυτανε[ίας· ἐκκλησία·]
κτλ.

Mounichion 22 = Prytany X 2[4] = 288th day.

Finally, the text of *I.G.*, II², 893*a*, belongs to the last month
of the year, and has been the subject of some discussion lately
not so much for its calendar as for the evidence it gives for the
place of meeting of the Ekklesia.[54] I have had the opportunity,
in May of 1958, to study this stone in Athens. The end of line 5
is indeed difficult to read, and it is easy to understand how one
could report no letters preserved in the last five letter-spaces.[55]
I was unable to see the letters ΛΗ of [βου]λή (or of [ἐκκ]λη-
[σία]), but before them the traces of ΕΚΚ of ἐκκλησία can, in
fact, still be distinguished, albeit faintly. The preserved traces
are Γ Ι Ι. So the word was not [βου]λή, but ἐ̣κ̣κλη[σία], and the
text of the opening lines may be read and restored as follows:

*I.G.*, II², 893*a*

*a.* 188/7 *a.*

[θ                    ε]              ο                    ι
['Επὶ Συμμάχου ἄρχο]ν[το]ς [ἐ]πὶ τῆς 'Αντιοχίδος δω
[δεκάτης πρυτανείας ἧι 'Αρ]χικλῆς Θεοδώρου Θορί
[κιος ἐγραμμάτευεν· Σκιρ]οφοριῶνος ἔκτει ἐπὶ ᵛ
5 [δέκα, ἑβδόμει καὶ δεκάτ]ει τῆς πρυτανείας· ἐ̣κ̣κλη

---

[54] B. D. Meritt, *A.J.P.*, LXXVIII, 1957, pp. 375-381, with a drawing on p. 377
and text on p. 378.
[55] Cf. *A.J.P.*, LXXVIII, 1957, p. 379 note 12.

[σία σύγκλητος ἐν τῶι] θεάτρωι μεταχθε[ῖ]σα ἐκ ᵛ
[Παναθηναϊκοῦ σταδίο]υ· τῶν προέδρων – – κτλ. –

The restoration in lines 5-6 is one which I had previously rejected because of the absence of any other evidence for a meeting of the Ekklesia in the stadion; but now that the text depends on a reading rather than on an outright restoration—so far as the Ekklesia is concerned—it seems best to accept it (with whatever reservations are necessary for the still restored portions) as new evidence for possible places of assembly. The calendar equation is:

Skirophorion 16 = Prytany XII [17] = 341st day.

This falls into place in the scheme of months and prytanies outlined above on p. 156.

## VI

Another example of forward count with μετ᾽ εἰκάδας is probably to be found in 178/7. Here Dinsmoor so interpreted the text of *Hesperia*, Suppl. I, pp. 120-121 (64) as to yield a normal ordinary calendar year.[56] The inscription has two equations:

Lines 2-3:

Πυανοψιῶνος ἐνάτει μετ᾽ εἰκάδας, τριακοστεῖ τῆς πρυ[τανε]ίας

Pyanepsion 29 = Prytany IV 30 = 118th day

Lines 28-29:

Μαιμακτηριῶνος ἔκτει ἱσταμένου, δεκάτει τῆς πρ[υ]τ[α]νείας

Maimakterion ⟨9⟩[57] = Prytany V 10 = 128th day.

[56] W. B. Dinsmoor, *Athenian Archon List*, p. 238.

[57] The assumption must be made that ἔκτει was cut on the stone by mistake for ἐνάτει; if it seems extreme to suppose such an error, one can show that a parallel exists in the cutting of *I.G.*, II², 1028B (line 67) [cf. *I.G.*, II², 1028A (line 2) for the correct reading].

Months and prytanies were probably arranged as follows:

Months       30   30   29   30   29 — — — — etc. — — — — — — — = 355
Prytanies   29   30   29   30 — — — — — — — etc. — — — — — — — = 355.

### VII

Mention should be made here of an equation from the archonship of Alexis (173/2), for which the text as restored by
Stamires (*Hesperia*, XXVI, 1957, p. 39) has been held by Pritchett (*B.C.H.*, LXXXI, 1957, p. 279 note 5) to be "surely incorrect." Stamires has restored two texts as of the same day (which
I consider highly probable), but he has given the dates only
by the calendar κατὰ θεόν and by the prytany. Normally, the
date κατ' ἄρχοντα preceded the date κατὰ θεόν, but there is one
text known in which a date κατ' ἄρχοντα followed the prytany
date (*Hesperia*, XVII, 1948, pp. 25-26). It did not inevitably,
therefore, precede the date κατὰ θεόν. The assumption made by
Stamires in this instance has been that the date κατ' ἄρχοντα was
omitted entirely: possibly, I should have said, incorrectly, but
not surely so. A substitute restoration can be made by which
Mounichion 11 κατ' ἄρχοντα is equated with Mounichion 12
κατὰ θεόν, which in turn equals Prytany X 20. But the calendar
upon which these equations could be posited is the same as that
implied by Stamires.

The other equations of this year are given in the text published
in *Hesperia*, XXVI, 1957, pp. 33-35 (6):

Metageitnion 2[1] = Prytany [II] 1[9] = 51st day
Metageitnion II 8  = Prytany [III] 4   = 68th day.

Months and prytanies within the year were arranged as
follows:

$$\overset{+1-1}{}$$

Months     30  30  29  30  29  30  29  30  29  30  29  30  29 = 384
Prytanies  32  32  32  32  32  32  32  32  32  32  32  32 = 384.

VIII

The year of Antigenes (171/0) was intercalary, but the two equations offered by the ephebic text now published in *Hesperia*, XV, 1946, pp. 198-201 (40) are still a problem. The second of the two equations (lines 45-46) is normal: ['Ελα]φηβολιῶνος ἐ[ν]ά[τε]ι ἱσταμένου, [ὀγ]δό[ει καὶ δεκάτει] τῆς πρυτα[νείας].

Elaphebolion 9 = Prytany [IX] [1]8 = 274th day.

The earlier of the two equations (lines 4-5) was read by me in 1946 as Πυανοψιῶνος [ἔ]ν[ει καὶ νέαι], ἑβδόμει καὶ δεκάτει τῆς πρυτανεία[ς].

Pyanepsion 29/30 = Prytany IV 17 = 118th day.

This gives five days too many for the first three prytanies of the year, an irregularity which would only be made worse by assuming that the archon had intercalated extra days in the festival calendar before the end of Pyanepsion. Pritchett and Neugebauer have suggested that in the date by month my reading [ἔ]ν[ει καὶ νέαι], which they could not verify from a squeeze, is incorrect, and they have proposed [ἐνάτει ἐπὶ δέκα], which is early enough to permit the assumption that days had been intercalated (irregularly) before it.[58]

I examined the stone again in Athens in 1958. The letters of Πυανοψιῶνος are quite clear down through the omega, though psi and iota are partly abraded. The omega is directly beneath the initial epsilon of 'Ερεχθηίδος in line 2. The nu following omega is largely lost, omikron is clear, and I confess that I cannot

---

[58] They have made other observations as well (*Calendars*, pp. 76-77): "The last letters of lines 4 and 5 which we can align with certainty are the first nu of Πυανοψιῶνος and the tau of πρυτανείας. To the right of the tau in line 5 were inscribed 17½ letters, iota being counted as half a letter. If prytanies and civil months were running uniformly, the civil date, which is equated with Prytany IV 17, should be restored as ἕκτει μετ' εἰκάδας. However, this restoration would require a space of 20½ letters and is probably too long. The phrase ἕκτει ἀπιόντος is a possible restoration, for which there is the analogy of *I.G.*, II², 951 (167/6)."

now see the final sigma. But of the date within the month the space upon the stone of each letter is certain, even when there is no assurance of what a given letter may have been. The date contained eleven letters, ending with an iota at the edge of the stone. This final letter can be distinguished, and before it was an alpha, the right side stroke discernible. Earlier than this the word καί is probable. If I were to read the text now afresh I should transcribe the date as Πυανοψιῶνο[ς ἔ]ν[ηι] καὶ [νέ]αι, with complete confidence in the accuracy of the record, however much the doubt about any individual letter. The other restorations that have been offered recently (ἐνάτει ἐπὶ δέκα or ἔκτει ἀπιόντος) are too long by two letters and do not match the traces on the stone, particularly the final alpha iota.

But this still leaves the calendar problem. I would suggest, in order to avoid the assumption of irregularly long prytanies, that the scribe (or copyist) had in his notation of the date by prytany the numeral ΔΔII which he mistakenly cut on the stone as ἑβδόμει καὶ δεκάτει.[59] In fact, the last day of Pyanepsion, in a regular intercalary year, should have fallen on the 22nd day of the fourth prytany.

IX

It is noticeable that a high percentage of those irregularities in which three or four or more days were added to the festival calendar occurred in months where they may be associated quite definitely, or most probably, with intentional postponements of the major Athenian festivals, the Panathenaia and the City

---

[59] See above, pp. 93, 112. If days could be omitted from the festival calendar, making the date κατ᾽ ἄρχοντα later than the date κατὰ θεόν (as suggested below, p. 206 note 11), then one might assume here that the equation is correct as it stands upon the stone and that five days were subtracted before the last day of Pyanepsion (ἔνη καὶ νέα) with compensation perhaps in the addition soon thereafter of five ἐμβόλιμοι ἡμέραι. The last day of Pyanepsion would then have been the 113th day of the year and properly equated with Prytany IV, 17.

Dionysia.⁶⁰ This was true of the year 295/4,⁶¹ it was true of 271/0,⁶² and of 244/3.⁶³ It was probably true in 228/7.⁶⁴ It was true in 196/5 and in 186/5.⁶⁵ It was probably true also of 160/59, which is represented by two texts, one from Athens and one from Delos. The inscriptions are:

*I.G.*, II², 953

a. 160/59 a.                                                   NON-ΣTOIX.

['Ε]πὶ Τυχάνδρου ἄρχοντος ἐπὶ τῆς 'Ακαμ[αντίδος ἔκτης πρυ]
[τ]ανείας ἧι Σωσιγ[έ]νης Μενεκράτου Μαρ[αθώνιος ἐγραμμάτευ]
[εν·] Ποσιδεῶνος δευτέραι μετ᾽ εἰκάδας, ἐ[νάτηι τῆς πρυτανεί]
[ας· ἐ]κκλησία κυρία ἐν τῶι θεάτρωι· – – [– – – κτλ. – – – –]

*Inscr. Délos*, 1497 bis

a. 160/59 a.                                                   NON-ΣTOIX.

[θ              ε]              ο              [ι]
['Επὶ Τυχάνδρου ἄρχοντ]ος ἐπὶ τῆς Οἰνείδος ἐνάτης {π}
[πρυτανείας ἧι Σωσιγέ]νης Μενεκράτου Μαραθώνιος ἔγραμ
[μάτευεν· 'Ελαφηβολιῶνο]ς τετράδι ἱσταμένου, ἐνάτει καὶ
5   [δεκάτει τῆς πρυτανεί]ας· ἐκκλησία κυρία ἐν τῶι θεάτρωι·
κτλ.

The first equation

Posideon 22 = Prytany [VI 9]

is normal for an intercalary year, if the count μετ᾽ εἰκάδας was forward. The second equation

[Elaphebolion] 4 = Prytany IX [1]9

⁶⁰ In 256/5 and in 248/7 there are indications of the intercalation of extra days (2 or 3, and 1 or 2, respectively) which cannot be associated with the major festivals. See Pritchett and Neugebauer, *Calendars*, pp. 72-73.
⁶¹ See above, pp. 26-33, 151.
⁶² See above, pp. 151-152.
⁶³ See above, pp. 146-148.
⁶⁴ See Pritchett and Neugebauer, *Calendars*, pp. 70, 73 with note 14.
⁶⁵ See Pritchett and Neugebauer, *Calendars*, pp. 75-76.

is not normal for either an intercalary or an ordinary year. The 19th day of the ninth prytany (275th of the year) should have been equated with Elaphebolion 9. In other words, there must have been an intercalation of five extra days in Elaphebolion,[66] again seemingly for the purpose of delaying the City Dionysia. These days could then have been compensated by omissions late in the month. The months and prytanies of the year may be arranged as follows:

$$\text{Months} \quad 29 \ 30 \ 29 \ 30 \ 29 \ 30 \ 30 \ 29 \ 30 \ \overset{+5-5}{29} \ 30 \ 29 \ 30 = 384$$
$$\text{Prytanies} \quad 32 \ 32 \ 32 \ 32 \ 32 \ 32 \ 32 \ 32 \ 32 \ 32 \ 32 \ 32 = 384.$$

Pritchett and Neugebauer, who think a forward count with μετ᾽ εἰκάδας not permissible,[67] restore ϵ[ἰκοστεῖ] instead of ἐ[νάτει] in line 3 of I.G., II², 953, and so obtain the equation

Posideon 29 = Prytany [VI 20] = 180th day.

Since the 180th day should have fallen on Posideon II 3, they assume an intercalation of four days before Posideon 29 (backward count) which created an irregularity in the calendar that was perpetuated throughout the rest of Posideon, all of Posideon II, all of Gamelion, all of Anthesterion, and at least until the latter part of Elaphebolion.[68] In a calendar where they suppose the months to have been regulated by continuous observation of the new moon this seems to me absurd, and I still prefer to understand the count μετ᾽ εἰκάδας in this inscription as forward, and to localize the calendar irregularity in Elaphebolion, associating it solely with a delay in the celebration of the City Dionysia and violating no date of new moon for any month anywhere in the year.

---

[66] In the light of this and of the examples earlier cited, one wonders whether perhaps an extra span of five days constituted a "normal" delay.

[67] Even Dinsmoor approves their insistence here on backward count. Cf. Amer. Hist. Rev., LIV, 1948/9, p. 337.

[68] Pritchett and Neugebauer, Calendars, pp. 28, 75, 76.

Another dislocation in the festival calendar which may be associated with the City Dionysia occurs in the year 164/3, now known to belong to the archonship of Euergetes.[69] In the seventh and eighth prytanies, in the months of Gamelion and Anthesterion, prytanies and months were in perfect accord and testify to an ordinary year without irregularities. There are two equations:

Hesperia, XXVI, 1957, p. 75 (cf. Hesperia,
Suppl. I, no. 79)

a. 164/3 a.                                  ΣΤΟΙΧ. 37

['Επὶ Εὐεργέτου ἄρχοντος] ἐπὶ τῆς 'Ερεχ[θεῖδος ᵛ]
[ἑβδόμης πρυτανείας] ἧι Δ]ιονυσόδωρος [Φιλίπ ᵛ]
[που Κεφαλῆθεν ἐγραμμάτε]νεν· δήμου ψη[φίσμα ᵛ]
[τα· Γαμηλιῶνος δεκάτει ὑστ]έραι, μιᾶι κ[αὶ εἰκο]
5   [στεῖ τῆς πρυτανείας· ἐκκλη]σία ἐμ Πει[ραιεῖ ᵛᵛ]
κτλ.

[Gamelion] 21 = Prytany [VII] [2]1 = 198th day

Hesperia, XXVI, 1957, p. 76 (cf. Hesperia,
Suppl. I, no. 79)

a. 164/3 a.                                  ΣΤΟΙΧ. 37

['Επὶ Εὐεργέτου ἄρχο]ντος ἐπὶ τῆς Λεωντ[ίδος ὁ ᵛ]
35   [γδόης πρυτανείας] ἧι Διονυσόδωρος Φι[λίππου]
[Κεφαλῆθεν ἐγραμ]μάτευεν· βουλῆς ψηφί[σματα ᵛ]
['Ανθεστηριῶνος τ]ετράδι ἱσταμένου, τε[τάρτηι]
[τῆς] πρυταν[είας· β]ουλὴ ἐμ βουλευτηρίω[ι· – ]
κτλ.

[Anthesterion] 4 = Prytany [VIII] 4 = 210th day.

In addition to these two equations there is now a third from the ninth prytany and the month of Elaphebolion:

⁶⁹ Hesperia, XXVI, 1957, pp. 72-77.

Hesperia, XXVI, 1957, pp. 72-73 (22)

a. 164/3 a.                                     ΝΟΝ-ΣΤΟΙΧ.

Ἐπὶ Εὐεργέτου ἄρχοντος ἐπὶ τῆς Ἱπποθωντίδος ἐνάτης πρυτ[α]

νείας ἧι Διονυσόδωρος Φιλίππου Κεφαλῆθεν ἐγραμμάτευε[ν·]

Ἐλαφηβολιῶνος ἐνάτει ἐπὶ δέκα, κατὰ θεὸν δὲ δεκάτει ὑστέ[ραι,]

δευτέραι καὶ εἰκοστεῖ τῆς πρυτανείας· ἐκκλησία ἐμ Πειρ[αιεῖ·]

κτλ.

Elaphebolion 19 ⟨κατ' ἄρχοντα⟩ = Elaphebolion 21 κατὰ θεόν =
Prytany IX 22 = 257th day.

Here the true date κατὰ θεόν corresponds most nearly to the
date by prytany, and it is evident that the date κατ' ἄρχοντα has
been retarded by the intercalation of two days, presumably
before the Dionysia, and presumably also destined to be corrected by omissions in the near future, probably before the end
of Elaphebolion.[70]

In this instance the extent of dislocation in the festival calendar
and at least part of the known time of it are indicated by the
designation of the true date as κατὰ θεόν. This example of double
dating in the festival calendar may now be added to those assembled by Pritchett and Neugebauer in 1947.[71] We add also two
texts from the archonship of Alexis in 173/2 (see above, p.
159) and two texts from the archonship of Theodotos
(95/4) published in Hesperia, XVII, 1948, pp. 25-26 (12).[72]
But we must subtract from the total number the earliest of all the
examples claimed (I.G., II², 861), which—in spite of Koehler's
suggestion: sumendum est v. 3. 4 diem mensis bis, κατ' ἄρχοντα

[70] The divergence of the date κατὰ θεόν from the date by prytany affords an
example (see above, p. 135) to show that ordinary years in the time of the
twelve phylai did not regularly have the days of months and prytanies in
one-to-one correspondence.

[71] Calendars, p. 15. The text there cited as Agora I 984 (unpublished) has
since appeared in Hesperia, XVI, 1947, p. 164 (64).

[72] See Pritchett, B.C.H., LXXXI, 1957, p. 279 note 5.

et κατὰ θεόν notatum fuisse, ut spatia expleantur—cannot be restored with such double dating.

The opening lines of this inscription are to be restored as follows:

*I.G.*, II², 861

*ante fin. saec.* III *a.*                                            *ca.* 47

['Επὶ – – – – – ἄρχοντος ἐπὶ τῆς – – – – πρ]ώτης πρυ

[τανείας ἧι – – – – – – – – – – – – – – – – – –]ς ἐγραμμάτε[υ]

[εν· δήμου ψηφίσματα· Ἑκατομβαιῶνος – – – με]τ᾽ εἰκάδας τ[ε

[τάρτει ἐμβολίμωι· – – – – – καὶ εἰκοστεῖ τῆς π]ρυτανείας·

5  [ἐκκλησία ἐν τῶι θεάτρωι· τῶν προέδρων ἐπεψήφιζ]εν Ἀπολλώ

[νιος (?) – – – – – – καὶ συμπρόεδροι]

[ἔδοξεν τῶι δήμωι]

κτλ.

While no date κατὰ θεόν is given, the text none the less bears witness to an irregularity in the calendar whereby four days were intercalated in the early twenties of the month Hekatombaion, apparently with a view to postponing the celebration of the Panathenaia. However restored, therefore, the date by prytany must be four days in advance of the date by month (e.g., ὀγδόει μετ᾽ εἰκάδας [23rd] and ἑβδόμει καὶ εἰκοστεῖ [27th]).

# CHAPTER VII

## THE THIRTEEN PHYLAI

From 223 to 201 B.C. the conciliar year in Athens was divided amongst thirteen phylai. In an intercalary year of thirteen months in the festival calendar, therefore, the prytanies and months must have been fairly evenly matched. This was surely so if the prytanies alternated between 29 and 30 days, like the months, but discrepancies of several days might arise if, as sometimes in the ordinary years when there were twelve phylai, the short or the long prytanies were grouped together either at the beginning or at the end of the year.

It has been assumed, recently, that in ordinary years, where the prytanies can have no close continuous correspondence with the months, each conciliar year was made up of three prytanies of 28 days followed by ten prytanies of 27 days.[1] The assumption is based, as its authors claim, on the pattern employed in the fourth century. I prefer to let the pattern of the third century speak for itself and then to use it (by analogy) to indicate that even in the fourth century the so-called Aristotelian rule was not rigidly applied.

In 203/2, for example, there is a text which indicates that the first prytany had only 27 days:

*I.G.*, II², 915 + *Hesperia*, Suppl. I, p. 89 (40) + *Hesperia*, XVII, 1948, pp. 15-16 (6)

a. 203/2 a.                       NON-ΣTOIX.

30   Ἐπὶ Προξενίδου ἄρχοντος ἐπὶ τῆς Ἱπποθωντίδος δευτέρα[ς πρυ]
τανείας ἧι Εὔβουλος Εὐβουλίδ[ο]υ Αἰξωνεὺς ἐγραμμάτευ[εν·]

---

[1] Pritchett and Neugebauer, *Calendars*, p. 89.

167

Μεταγειτνιῶνος δευτέραι ἱσταμένου, πέμπτηι τῆς πρυ[τα]
νείας· βουλὴ ἐμ βουλευτηρίωι – – – – – κτλ. – – – – –

Metageitnion 2 = Prytany II 5 = 32nd day.

Here Hekatombaion had 30 days, and the first prytany had
27 days. One must assume the intercalation of an extra day
somewhere in Hekatombaion, without compensating omission,
in order to push the first prytany up to the hypothetically
desired complement of 28 days.[2] But this assumption of irregularity is quite unnecessary. There is now another text, not available when Pritchett and Neugebauer made their study, belonging to the year 214/3, which also indicates that the first prytany
of the year had only 27 days:

Hesperia, XXIII, 1954, pp. 236-237 (7)

a. 214/3 a. NON-ΣΤΟΙΧ.

['Επὶ Εὐφιλ]ήτου ἄρχοντος ἐπὶ τῆς ⟦['Αν[τιγονίδος]]⟧ δευτέρας
πρυ]
[τανεί]ας ἧι 'Αρίστων Θεοδώρου 'Ραμν[ούσιος ἐγραμμάτευεν·]
[ψη]φίσματα δήμου· Μεταγειτνιῶνος ἔκ[τει ἐπὶ δέκα δευτέ]
ραι ἐμβολίμωι, μιᾶι καὶ εἰκοστεῖ τῆς πρυτα[νείας· ἐκκλησία]
5 ἐν τῶι θεάτρωι· – – – – – – – – κτλ. – – – [– – – – – – –]

Metageitnion [1]6 twice intercalated = Prytany [II] 21 = 48th day.

I have earlier suggested that the assumption of still another
intercalated day somewhere before Metageitnion 16 would
enable one to give the first prytany a total of 28 days.[3] But this
is quite gratuitous, and in view of the evidence also from 203/2
I believe that one should acknowledge prytanies of 27 days with
no assumption of attendant irregularity.

There is no certain evidence about the prytanies in the pre-

---

[2] This is, in fact, the assumption made by Pritchett and Neugebauer,
Calendars, pp. 91, 93.
[3] Hesperia, XXIII, 1954, p. 238.

served inscriptions of 219/8. The better calendar equation is in the text of an inscription from the Agora first published in 1933 and republished, with improvements, in 1942: [4]

Hesperia, XI, 1942, pp. 298-299 (59)

a. 219/8 a.                                NON-ΣΤΟΙΧ.

['E]πὶ Χαιρεφῶντος ἄρχοντ[ος ἐπὶ τῆς — — — —]
[τ]ρίτης πρυτανείας ἧι Φ[— — — — — — — — — —]
[Κυ]δαντίδης ἐγραμμάτ[ευεν· Βοηδρομιῶνος]
[ἐν]δεκάτει, πέμπτηι κ[αὶ δεκάτηι τῆς πρυ]
5    [ταν]είας· ἐκκλησία· — [— — κτλ. — — — — —]

[Boedromion] 1[1] = Prytany III [1]5 = 70th day.

It is quite possible that this equation is correct, with the first prytany of 27 days, as in 214/3 and 203/2, in which case the second prytany had 28 days. But the order may have been reversed. Or, if the date within the month is restored as the 12th then it may be supposed that the first two prytanies each had 28 days. In any of these various combinations there is no evidence of irregularity or of any tampering with the festival calendar. Another decree of the archonship of Chairephon (Hesperia, XXIX, 1960, p. 76, no. 153) is not well enough preserved to be helpful for the calendar.

The same indifference as to which prytanies in the ordinary year had 27 days and which had 28 is exhibited also in the one preserved equation of the year 218/7:

I.G., II², 843

a. 218/7 a.                                ΣΤΟΙΧ. 39

[θ          ε]              ο              [ι]
['Επὶ Καλλι . . .⁵. . ἄρχ]οντος ἐπὶ τῆς Αἰγεῖδος δωδε
[κάτης πρυτανείας ἧ]ι 'Αριστοτέλης Θεαινέτου Κε

─────────────────────

⁴ The equation given with this text is that restored in Hesperia, XI, 1942, pp. 298-299.

[φαλῆθεν ἐγραμμάτε]νεν· Θαργηλιῶνος δεκάτει <sup>vv</sup>
[ἱσταμένου, ἕκτει τῆς] πρυτανείας· ἐκκλησία – –
κτλ.

Thargelion 10 = Prytany XII [6] [5] = 305th day.

It does not matter whether Thargelion was full and Skirophorion hollow or *vice versa*: the year still had 49 days to run, and the last two prytanies had 27 and 28 days (or 28 and 27). If perchance the last two months of the year were both of 30 days (the year being an ordinary year of 355 days) then the last two prytanies must have had 28 days each. But it is evident that the year did not commence with three prytanies of 28 days, to be followed by ten prytanies of 27 days. Indeed, none of the evidence from the years between 223 and 201 shows that the Athenians patterned their prytany calendar of the late third century on the analogy of Aristotle's schematic definition of the calendar of the fourth century. On the contrary, the evidence favors an unpredictable succession of 27- and 28-day prytanies.

This can be further demonstrated by reference to an inscription of the archonship of Ankylos (208/7) which has been universally misunderstood. As published first by Sterling Dow, the text was interpreted on the assumption that an entire line, and more, had been omitted from the prescript.

*Hesperia,* Suppl. I, p. 86 (38) [6]

a. 208/7 a.                              NON-ΣΤΟΙΧ.

1 Ἐπὶ Ἀγκύλου ἄρχοντος ἐπὶ τῆς Πανδιονίδος δεκ[ά]
⟨της πρυτανείας ἧι – – – – – – – – – ἐγραμμάτευεν·⟩

_____

[5] The restoration is guaranteed by the stoichedon order.

[6] Dow, *Hesperia,* Suppl. I, p. 88, writes: "It is plausible to guess that the mason's eye jumped from the numeral defining the place of the prytany to the (similar) numeral denoting the day of the month. The year was evidently ordinary, with a discrepancy in the tenth prytany of three days between month and prytany dates."

⟨Μουνιχιῶνος ? δεκά⟩
2  τει ὑστέραι, τετάρτει καὶ εἰκοστεῖ τῆς πρυτανε[ίας· ἐκ]
3  κλησία ἐμ Π⟨ει⟩ραεῖ· – – – – – – κτλ. – – – – – – –

⟨Mounichion ?⟩ 21 = Prytany X 24

Dow considered the year surely ordinary.[7] But Pritchett and
Neugebauer, though accepting Dow's reconstruction of the text,
claim the year as surely intercalary and suggest moving the
archon Ankylos to some date other than 208/7.[8] It would have
been more satisfactory, from the beginning, not to have posited
an error in the transmitted text. It is quite coherent and intelligible as it stands, and should be read as follows:

*Hesperia*, Suppl. I, p. 86 (38)

a. 208/7 a.                                    NON-ΣΤΟΙΧ.

Ἐπὶ Ἀγκύλου ἄρχοντος ἐπὶ τῆς Πανδιονίδος· δεκ[ά]
τει ὑστέραι, τετάρτει καὶ εἰκοστεῖ τῆς πρυτανε[ίας· ἐκ]
κλησία ἐμ Π⟨ει⟩ραεῖ· – – – – – – κτλ. – – – – – –

There was clearly an omission, but it may have been intentional and need not be ascribed to error or oversight.[9] There
is no evidence that Pandionis held the tenth prytany, or that the
month was Mounichion. If the year be taken as ordinary the
calendar equation is correct for the second prytany and the
month of Metageitnion:

⟨Metageitnion⟩ 21 = Prytany ⟨II⟩ 24 = 51st day.

Again it appears that the first prytany may have had only 27
days. This inscription may, however, also be restored as of the

[7] As also, at different times, have Dinsmoor, Pritchett, and I. See Pritchett
and Neugebauer, *Calendars*, p. 90 with note 2. Dinsmoor (*Hesperia*, XXIII,
1954, p. 316) followed Pritchett and Neugebauer in making the year of Ankylos
intercalary, and he dated the archonship in 207/6.
[8] Pritchett and Neugebauer, *Calendars*, pp. 90-91.
[9] For the prytany without ordinal number, see, for example, the text of
*I.G.*, I², 304B, as given in Meritt, *Athenian Financial Documents*, p. 120, line 71.

third prytany and the month of Boedromion:

⟨Boedromion⟩ 21 = Prytany ⟨III⟩ 24 = 80th day.

In this case the first two prytanies each had 28 days. The evidence is ambiguous, and the text should not be added to those of 219/8, 218/7, and 203/2, to show that an initial prytany of 27 days was a normal phenomenon.

There is further evidence about the calendar in another year of the thirteen phylai (222/1) for which there are two equations in the text of *I.G.*, II², 848, as well as a reference in *I.G.*, II², 844, to the fact that Anthesterion was intercalated. This latter would seem, on the face of it, to make the year an intercalary year, but there is no corresponding date by prytany during the intercalated month. This leads to uncertainty, and the extra month has been interpreted, in recent studies,[10] as a month intercalated κατ' ἄρχοντα, it being assumed that an equivalent number of days were subsequently omitted and that the whole year, κατ' ἄρχοντα as well as κατὰ θεόν, was an ordinary year of twelve months. The earlier assumption had been that the year started as ordinary, that the decision to intercalate was made late, and that the prytanies during the last half of the year were correspondingly lengthened.[11] Without a prytany date in the latter part of the year this issue could hardly be settled with certainty.

No evidence exists for the dogma that the character of a festival year could not, if the Athenian demos so wished, be changed while in course from ordinary to intercalary. A prytany date after the late intercalation would here disclose (as it did in 307/6; see below, pp. 176-178) whether the year was intercalary κατὰ θεόν as well as κατ' ἄρχοντα.

But the evidence of *I.G.*, II², 848, shows that at least the year

---

[10] Pritchett and Neugebauer, *Calendars*, pp. 91-92.
[11] W. B. Dinsmoor, *Archons*, p. 217; *Athenian Archon List*, pp. 232-233; Pritchett and Meritt, *Chronology*, p. xxiv.

began as ordinary. The equation in lines 1-3, unfortunately, is not preserved, except to show that some part of Boedromion fell in the twenties of the third prytany. The equation in lines 27-29 reads as follows: [12]

**I.G., II², 848 (lines 27-29)**

*a.* 222/1 *a.*                                              NON-ΣΤΟΙΧ.

Ἐπ' Ἀρχελάου ἄρχ[ον]τος [ἐπὶ τῆς] Αἰαντίδος τ[ετ]άρτης πρυτα
νείας ἧι Μόσχος Μοσ[χίωνος Ἀ]γκυλῆθεν ἐγραμμάτευεν· Βο
ηδρομιῶνος ἐβδ[ό]μει [μετ' ε]ἰκ[άδα]ς, [τρ]ίτει τῆς πρυτανείας·

Boedromion 27 = Prytany IV 3 = 86th day.

The count with μετ' εἰκάδας is forward, and of the first three prytanies two had 28 days and one had 27. There is no irregularity here in such a disposition of prytanies.[13] Yet the year 222/1, which we have found to be an ordinary year and at the same time an intercalary year with an extra Anthesterion, poses an interesting calendar problem. It may be instructive to learn how the following year began, and to see what its calendar character, in turn, may have been.[14] The year belonged to the

---

[12] Pritchett and Neugebauer have given it as their opinion (*Calendars*, p. 91 note 8) that all the letters in the date by month should be read as dotted letters, i. e., as doubtful. There is, in my opinion, not the slightest doubt about the date. I examined the stone again in Athens in 1958 and read with certainty, as I had on the squeeze, the letters here given in the text (cf. *Hesperia*, IV, 1935, p. 557). I have placed a dot under initial epsilon because I can see only the vertical and top horizontal strokes and a dot under the next epsilon because again the letter is not entirely preserved; but in neither case is there doubt about what the letter was.

[13] See above, p. 170. Pritchett and Neugebauer, *Calendars*, pp. 91-92, write: "According to the theory of regular prytany lengths, the third day of the fourth prytany should fall on the 87th day of an ordinary year, whereas Boedromion 24 would be approximately the 83rd day of an undisturbed civil year. We, therefore, conclude that extra days had been intercalated into the early months of the civil calendar and disagree with our predecessors in their determinations that the equation is evidence for irregularity in the length of the first prytanies and that the count with ἑβδόμει μετ' εἰκάδας was forward." Cf. also *Calendars*, p. 28.

[14] There is no discussion of, or reference to, the calendar of 221/0 in Pritchett and Neugebauer, *Calendars*.

archonship of Thrasyphon, and has one preserved calendar equation:

I.G., II², 839

a. 221/0 a.                              NON-ΣΤΟΙΧ.

(lines 1-5 not repeated here)

6  Ἐπὶ Θρασυφῶντος ἄρχοντος [ἐπὶ τῆς Πανδι]
   ονίδος ἕκτης πρυτανείας ἧι [— — ᶜᵃ· ¹⁰ — — — —]
   του Παιανιεὺς ἐγραμμάτε[νεν· δήμου ψη]
9  φίσματα· Μαιμακτηριῶνος [ἔνηι καὶ νέαι,]
   ἕκτει καὶ δεκάτει τῆς πρυτ[ανείας· — κτλ. —]

Kirchner's comment on the year was "Annus est communis ab Hecatombaeone solido incipiens, quo prytaniae I-III 26 dies, IV-VII 27, VIII-XIII 28 tenebant. Iam dies 30 Maemacterionis respondet diei 16 prytaniae sextae. Lacunae v. 9 nihil aliud nisi supplementum ἔνηι καὶ νέαι accomodatum esse iam vidit Reusch; sane cum ille sub finem s. III tredecim exstitisse tribus etiamtum ignoraret, quomodo dies 16 prytaniae sextae incideret in Maemacterionem expedire non potuit."

Surely, if the first three prytanies seemed to have only 26 days each, a calendar irregularity would have to be assumed, with extra intercalations early in the year in the festival calendar. But Reusch's observation that only the restoration ἔνηι καὶ νέαι is possible at the end of line 9 is incorrect. The restoration can and should be ἕκτει ἐπὶ δέκα, showing that Maimakterion 16 was equated with the 16th day of the prytany, and that the months and prytanies were in perfect accord.

But the prytany was the sixth, and Maimakterion was the fifth month. The normal explanation of this phenomenon is that some month earlier than Maimakterion had been intercalated.[15] This can, indeed, be the explanation here, and the

---

[15] For the frequency of this type of intercalation see Pritchett and Neugebauer, Calendars, p. 90.

difficulties both in 222/1 and in 221/0 can be resolved on the assumption that the intercalated month was the preceding Anthesterion, already attested κατ' ἄρχοντα, and an extra Hekatombaion κατὰ θεόν. The calendar of these two years, over the interval in question, is probably to be reconstructed as follows:

222/1 B.C.

| | |
|---|---|
| Anthesterion κατ' ἄρχοντα | = Anthesterion κατὰ θεόν |
| Anthesterion II κατ' ἄρχοντα | = Elaphebolion κατὰ θεόν |
| (I.G., II², 844, line 33) | |
| Elaphebolion κατ' ἄρχοντα | = Mounichion κατὰ θεόν |
| Mounichion κατ' ἄρχοντα | = Thargelion κατὰ θεόν |
| Thargelion κατ' ἄρχοντα | = Skirophorion κατὰ θεόν |

Here the conciliar year, equivalent to the festival year κατὰ θεόν, ended, and the new conciliar year 221/0 commenced.

| | |
|---|---|
| Skirophorion κατ' ἄρχοντα | = Hekatombaion κατὰ θεόν |
| Hekatombaion κατ' ἄρχοντα | = Hekatombaion II κατὰ θεόν |
| Metageitnion κατ' ἄρχοντα | = Metageitnion κατὰ θεόν |
| Boedromion κατ' ἄρχοντα | = Boedromion κατὰ θεόν |
| Pyanepsion κατ' ἄρχοντα | = Pyanepsion κατὰ θεόν |
| Maimakterion κατ' ἄρχοντα | = Maimakterion κατὰ θεόν |

The year 221/0 was an intercalary year κατὰ θεόν, with the months and prytanies in accord, and with the equation in I.G., II², 839:
Maimakterion [16] = Prytany VI 16 = 163rd or 164th day.

This apparent running over of one year, that had been made unduly long by an unexpected intercalation, into the beginning of the next was a device for getting the calendar back to normal after the introduced abnormality. In 222/1 and 221/0 it had the effect of making an intercalary year precede an ordinary year, both κατ' ἄρχοντα, while the same two years were respectively ordinary and intercalary κατὰ θεόν, the prytanies of the conciliar year being scaled to the festival dates κατὰ θεόν.

But this was not the only method of dealing with the problem. There was an irregular intercalation in 166/5, and yet the year ended (apparently) as it had begun, an ordinary year κατ' ἄρχοντα and κατὰ θεόν. Omissions of days before the end of the year compensated for the intercalation and the following year (ordinary) was not in any way affected and was not called upon to give help in making the calendar again normal. The year 166/5 was for a time intercalary κατ᾽ ἄρχοντα but rectified itself and ended as an ordinary year. The prytanies of the conciliar year were scaled to the festival dates κατὰ θεόν throughout.[16]

There was a third method of dealing with this problem. If a year began as ordinary and was given an unexpected intercalary month, that year (with its added month) might be carried on as intercalary both κατ᾽ ἄρχοντα and κατὰ θεόν. Under these circumstances the lengths of prytanies in the conciliar year were readjusted so as to absorb the days of the extra month within those prytanies that remained after the moment of intercalation. An example of this method is found in the year 307/6, and has already been described by Pritchett and Meritt in *Chronology*, pp. 12-21. They outlined a year in which the months and prytanies had the following lengths (*op. cit.*, p. 21):

Months    29   30   29   30   29   30   30   29   30   29   30   29   30 = 384
Prytanies   30   30   30   30   30   29   35   34   35   35   33   33 = 384.

There was, however, one calendar equation in which the date Γαμηλιῶνος δευτ[έ]ραι ἐ[μ]βολίμωι ὀγδόε[ι] μετ᾽ εἰκάδας ἡμερολεγδόν fell on the 21st day of the seventh prytany and in which the festival date was supposedly repeated (by intercalation) only once. We now know that the specification δευτ[έ]ραι ἐ[μ]βολίμωι means a second intercalation of the day named. Therefore, we may arrange the first six (not five) prytanies with 30 days each and achieve the equation of the epigraphical text (*I.G.*, II², 458) by allowing the month Gamelion to be hollow and counting the second intercalated Γαμηλιῶνος ὀγδόη μετ᾽ εἰκάδας as Gamelion 24 (the date ὀγδόη μετ᾽ εἰκάδας being the 22nd ἡμερολεγδόν in a hollow month).

I assume that the compensation was made before the end of

16 See below, pp. 183-184.

the month, so that Gamelion had, in all, only 29 days. Inasmuch as the year began as ordinary with six prytanies of 30 days each, it was obviously the intention (before the irregular intercalation) to end the year with six prytanies of 29 days each. But the decision to intercalate a second Gamelion,[17] made clearly during the first Gamelion, put an extra span of 30 days into the year, to be divided out amongst the last six prytanies. The regularity of prytany lengths could be maintained if five days were added to each of the prytanies from VII to XII; each of these prytanies, then, had 34 days rather than the anticipated 29 days. Moreover, this distribution can be made to square with the epigraphical evidence if the calendar equation in *I.G.*, II², 455, be read as

3  [ἴδος ἑνδεκάτης πρυτανείας ἧι Λυσίας Νοθί]ππου Διομ
4  [ειεὺς ἐγραμμάτευεν· Θαργηλιῶνος δευτέρα]ι ἱσταμέν
5  [ου, δεκάτει τῆς πρυτανείας· ἐκκλησία· τῶν πρ]οέδρων ἐπ

and the calendar equation in *I.G.*, II², 460, be read as

2  [ος δωδε]κάτης πρυ[τανείας ἧι Λυσίας Νοθίππο]
3  [υ Διομε]ιεὺς ἐγραμ[μάτευεν· Σκιροφοριῶνος τ]
4  [ρίτει] ἱσταμένου, ἑ[βδόμει τῆς πρυτανέας· ἐκκ][18]

and the calendar equation in *I.G.*, II², 461, be read as

4  [αμμάτευεν· Ἐλαφηβ]ολιῶνο[ς ἕκτει μ]
5  [ετ' εἰκάδας, ὀγδόει] τῆς πρυ[τανείας·]

The months and prytanies within the year were as follows:

Months     29 30 29 30 29 30 29 30 29 30 29 30  30 = 384
Prytanies  30 30 30 30 30 30 34 34 34 34 34 34 = 384.

and the equations derived from the preserved inscriptions are as follows:

---

[17] The evidence is in *I.G.*, II², 1487.
[18] For the spelling πρυτανέας, see Pritchett and Neugebauer, *Calendars*, p. 38.

Pritchett and Meritt, *Chronology*, p. 8

Hekatombaion [11] = Prytany [I 11] = 11th day

*I.G.*, II², 464

[Pyanepsion⟨30⟩] = Prytany [IV] 28 = 118th day

*I.G.*, II², 456

Maimakterion [29] = Prytany V 2 [7] = 147th day

*I.G.*, II², 458

Gamelion 22 (twice intercalated) = Gamelion 24
= Prytany VII 21 = 201st day

*I.G.*, II² 459

Anthesterion [20] = Prytany [IX] 8 = 256th day

*I.G.*, II², 461

Elaphebolion [25] = Prytany [X 8] = 290th day

*I.G.*, II², 455

[Thargelion 2] = Prytany [XI 10] = 326th day

*I.G.*, II², 460

[Skirophorion 3] = Prytany [XII 7] = 357th day.

In all cases the texts and commentary are to be read as in *Chronology*, pp. 12-21, except as noted here immediately above.[19]

This spreading out of extra days during the last prytanies of the year finds its contrast in the division of a few days among all the prytanies, as was the case in 296/5. When the tyranny of Lachares was overthrown in the spring of that year and democratic institutions re-established, the remaining days to the end of the festival year were distributed among the twelve prytanies so that each prytany had eight or nine days.[20]

---

[19] Pritchett and Neugebauer, *Calendars*, p. 69, "have attempted no reconstruction of the calendar of the year 307/6, in the course of which the number of prytanies was increased from 10 to 12." There is a misapprehension here about the number of the prytanies. Their number was already fixed as 12 before the year began (cf. Pritchett and Meritt, *Chronology*, p. 21).

[20] I agree with Unger that the reëstablishment took place immediately after the Dionysiac festival (*Philologus*, XXXVIII, 1879, pp. 445-446).

The evidence is in *I.G.*, II², 644:

*a.* 296/5 *a.*                    ΣΤΟΙΧ. 27

Ἐπὶ Νικίου ἄρχοντος ὑστέρ[ου ἐπὶ]
τῆς Ἀκαμ[α]ντίδος τετάρτης π[ρυτα]
[νε]ίας ἧι Ἀ[ν]τι[κρ]άτης Κρατίν[ου Ἀζ]
[ην]ι[εὺς ἐγραμμ]άτευε· Μουνιχ[ιῶ ᵛᵛ]
5  [ν]ος ἕκ[τηι ἐπὶ δέ]κ[α], ἑβδόμη[ι τῆς ᵛᵛ]
[π]ρυτα[νείας· ἐκκ]λη[σ]ία· τῶ[ν προέδρ]
[ων ἐ]πε[ψήφιζε ...] \P \Γ------

Mounichion 16 = Prytany IV 7 = 281st day.

The first day of the first prytany, in the re-established Council, must have fallen on, or very nearly on, Elaphebolion 14.

The inscription published as *I.G.*, II², 973, has been variously assigned, but now is generally agreed to belong to the year 204/3.[21] It shows the close correspondence between months and prytanies which is to be expected in an intercalary year in the time of the thirteen phylai:

*I.G.*, II², 973

*a.* 204/3 *a.*                    NON-ΣΤΟΙΧ.

   θ           ε           ο           [ί]
Ἐπὶ Ἀπολλοδώρου ἄρχοντος [ἐπὶ τῆς Πανδιο]
νίδος δευτέρας πρυτανεία[ς ἧι -------]
νος Ὀῆθεν ἐγραμμάτευεν· [Μεταγειτνιῶνος]
5  [ἑ]νδεκάτῃ, ἑνδεκάτει τ[ῆς πρυτανείας· ἐκκλη]
[σί]α κυρία ἐν τῶι θε[άτρωι· τῶν προέδρων ἐπε]
[ψή]φιζεν Ἀθη[ν]ογ[ένης Εὐνόμου Λευκονοεύς?]

[Metageitnion] 11 = Prytany II 11 = 40th or 41st day.

---

[21] See W. K. Pritchett, *Hesperia*, XVI, 1947, pp. 190-191.

# CHAPTER VIII

## COINS AND THE CALENDAR

The hope that historians have long entertained for establishing sequences of ordinary and intercalary years by the numismatic evidence of Athenian New Style coinage is moving closer to realization. Something must be said about this here, for it conditions the restoration of a number of epigraphical texts from the second century. Reference to the pioneer work of M. L. Kambanis in arranging the issues of New Style coinage by a study of sequences of dies was made by Pritchett and Neugebauer in their book on the calendar,[1] but except to note the opportunity for possibly valuable contributions from further numismatic study there was little that could be cited as valid evidence at that time. A summary of Kambanis's results was published by Margaret Thompson in 1941.[2] In the meantime she has continued the work begun by Kambanis and extended her examination of the coins to give the widest possible base for a study of die sequences.

In 1951 the American Numismatic Society acquired casts of the New Style tetradrachms which Kambanis had assembled, and a preliminary study was made by Miss Thompson in which she used originals, casts, or photographs of over 3000 coins.[3] In the same year she offered reasons for believing that the New Style coinage was begun in 196/5 B.C. after the proclamation of freedom for the Greek cities by Flamininus.[4] By 1955 a col-

---

[1] Pritchett and Neugebauer, *Calendars*, pp. 109-110.
[2] Margaret Thompson, *Hesperia*, X, 1941, p. 207 note 22.
[3] Margaret Thompson, "Workshops or Mines," *American Numismatic Society Museum Notes*, V, 1952, pp. 35-48.
[4] Margaret Thompson, "The Beginning of the Athenian New Style Coinage," *American Numismatic Society Museum Notes*, V, 1952, pp. 25-33. See also M.

lection of 6000 coins, casts, and photographs had been assembled
at the American Numismatic Society's museum, and a grant was
made from the Penrose research funds of the American Philosophical Society to enable Miss Thompson to complete her
study of the known private and public cabinets where Athenian
New Style coinage might be found. Visits to Glasgow, Athens,
and Istanbul were especially fruitful; many well-known collections were again studied; and Miss Thompson was able to report
that her final publication of Athenian New Style coinage could
go forward "in the confidence that all the major numismatic
collections presently accessible have been studied and that the
information assembled is as complete as one can hope to
make it."[5]

This work is now in press (at present writing), but I am
permitted to use the evidence of the die sequences studied by
Miss Thompson to say that intercalary years between 196 and
100 B.C. seem to have occurred in 184/3, 171/0, 170/69, 167/6,
162/1, 154/3, 137/6, 134/3, 125/4, 119/8, and 113/2 B.C. These
are the years from which coins have been preserved showing
the letter nu, the thirteenth letter, indicating a thirteenth month,
and hence evidence for an intercalary year. It is understood, of
course, that lack of a month-letter nu in other years does not
prove them ordinary; the evidence, if it existed, simply may not
have been preserved.

It remains to be seen how the numismatic tallies with the
epigraphic (and other) evidence.

There is no evidence for the year 184/3, though this has been
taken as ordinary[6] in the reconstruction of a hypothetical sequence for this and adjacent years. We now make 184/3 inter-

Thompson, "The Grain-Ear Drachms of Athens," *Centennial Publication of the American Numismatic Society* (1958), p. 653 note 2.
[5] *Year Book of the American Philosophical Society*, 1956, pp. 354-357.
[6] E.g., by Pritchett and Meritt, *Chronology*, p. xxvii.

calary, and by way of recompense make 183/2, for which also there is no other evidence, ordinary.

At first sight it is disconcerting to find that the numismatic evidence requires two intercalary years in succession in 171/0 and 170/69. But this is not impossible,[7] and there is no epigraphic evidence against it. Indeed, the archonship of Antigenes (171/0) is known to have been intercalary.[8] No calendar equation is preserved for the year 170/69, to which I now attribute the archon Aphrodisios.[9] Since the numismatic evidence calls for 170/69 to be intercalary, I so indicate it, and call attention to the compensation which this piling up of intercalary years finds in the succession of ordinary years immediately following. But first one should note that in all probability 168/7 was ordinary and 167/6 intercalary. The one equation of the year of Xenokles (168/7) is in the text of *I.G.*, II², 945: [10]

*I.G.*, II², 945

*a.* 168/7 *a.*                            NON-ΣTOIX.

θ              ε              ο              ι
Ἐπὶ Ξενοκλέους ἄρχοντος ἐπὶ τῆς Οἰνεῖδος δωδεκάτης πρυτα
νείας ἧι Σθενέδημος Ἀσκ⟨λ⟩ηπιάδου Τειθράσιος ἐγραμμάτευεν·
Σκιροφοριῶνος ἔνει καὶ νέαι, μιᾶι καὶ τριακοστεῖ τῆς πρυ[τανείας·]
κτλ.

Skirophorion 30 = Prytany XII 31 [11] = 355th day (last day).

[7] It is known that there were two intercalary years in succession once, at least, in the late fifth century (see below, p. 218), and there were probably two in succession in the third century (264/3 and 263/2; see below, p. 233). The last six years of the ninth Metonic cycle would "normally" have been OOIOIO (267/6–262/1). Actually, the intercalation which might have been expected in 265/4 seems to have been postponed to 264/3.

[8] See above, pp. 160-161.

[9] See below, pp. 198-199.

[10] Usually considered as surely intercalary: cf. Pritchett and Meritt, *Chronology*, p. xxviii; Pritchett and Neugebauer, *Calendars*, p. 75; B. D. Meritt, *Hesperia*, XXVI, 1957, p. 95.

[11] Pritchett and Neugebauer, *Calendars*, p. 78, allow a prytany of 31 days at the end of a festival year of 355 days, and we have seen above (pp. 140, 144-145) that prytanies of 31 days are attested elsewhere as well in ordinary years.

The year 167/6 is represented by a text from the Agora of Athens published by Sterling Dow:

Hesperia, Suppl. I, p. 135 (72)

a. 167/6 a.                                          NON-ΣTOIX.

Ἐπὶ Νικοσθένου ἄρχοντος ἐπὶ τῆς Οἰνεῖδος ἕκτης πρυτανεί

[ας ἧι – – – – – – <sup>ca. 23</sup> – – – – – –]ος ἐγραμμάτευεν· δήμου

[ψηφίσματα· Ποσιδεῶνος δευτέραι μετ᾽ εἰκάδας,] δευτέραι καὶ εἰ

[κοστεῖ τῆς πρυτανείας· ἐκκλησία ἐν τῶι θεάτρωι· ] τῶν – κτλ. –

Dow remarked that the year could be restored as ordinary, and it has been so interpreted.[12] But it can also be restored as intercalary, and in view of the numismatic evidence the intercalary interpretation is to be preferred:

Hesperia, Suppl. I, p. 135 (72)

a. 167/6 a.                                          NON-ΣTOIX.

Ἐπὶ Νικοσθένου ἄρχοντος ἐπὶ τῆς Οἰνεῖδος ἕκτης πρυτανεί

[ας ἧι – – – – – <sup>ca. 23</sup> – – – – – – –]ος ἐγραμμάτευεν· δήμου

[ψήφισμα· Ποσιδεῶνος ὑστέρου πέμπτει ἱσταμένου,] δευτέραι καὶ εἰ

[κοστεῖ τῆς πρυτανείας· ἐκκλησία ἐν τῶι θεάτρωι·] τῶν – κτλ. –

[Posideon II 5] = Prytany VI 22 = 182nd day.

With the year 166/5 we are confronted with one of the most serious anomalies in the festival calendar. Roughly speaking, Anthesterion κατ᾽ ἄρχοντα was equivalent to Elaphebolion κατὰ θεόν,[13] and Mounichion κατ᾽ ἄρχοντα was equivalent to Thargelion κατὰ θεόν.[14] But it is clear that the festival year κατὰ θεόν was ordinary,[15] and it is now equally clear that the three follow-

---

[12] Pritchett and Meritt, Chronology, p. xxix; Pritchett and Neugebauer, Calendars, p. 85.

[13] I.G., II², 946, as interpreted by Pritchett and Neugebauer, Calendars, p. 85 with note 25.

[14] I.G., II², 947, lines 10-11.

[15] See Hesperia, III, 1934, p. 21 (19) = Hesperia, Suppl. I, no. 73; I.G., II², 948; Pritchett and Meritt, Chronology, p. 127.

ing years 165/4, 164/3, and 163/2 were ordinary as well.[16] The apparent succession of five (or even four) ordinary years which I did not understand when I wrote in 1957 that "the calendar of this year needs further study" is now explicable in the light of the successive intercalary years of 171/0 and 170/69.[17]

After the ordinary years 166/5, 165/4, 164/3, and 163/2, it is only natural to assume that the year 162/1 was intercalary.[18] There is no direct epigraphical record, but this assumption is now confirmed by the numismatic evidence.

The next year indicated as intercalary by the numismatic evidence was 154/3. There is no epigraphical control. But the intercalation raises no problem, for 157/6 (archonship of Anthesterios) was surely intercalary[19] and 155/4 (archonship of Mnesitheos) was surely ordinary.[20]

I have recently had word (by letter) from Jacques Tréheux that the archon immediately preceding Anthesterios is named in a Delian inscription as Pyrrhos.[21] The discovery of a new name, always welcome, raises unexpected problems, for there now seem to be too many archon-names for the years available, and Aristaichmos can no longer be assigned to 158/7. The only date for him, in fact, is 159/8, long advocated by P. Roussel;[22] but this in turn displaces Demetrios, whom I had thought (only within the last few years) to have placed with some assurance in that year. The problem of the archons of these years resolves itself, then, into the problem of how Demetrios can be reassigned.[23]

---

[16] See *Hesperia*, XXVI, 1957, p. 95.
[17] See above, p. 182.
[18] See *Hesperia*, XXVI, 1957, p. 95; Pritchett and Meritt, *Chronology*, p. xxix.
[19] See J. H. Kent, *Hesperia*, XVI, 1947, p. 224.
[20] Pritchett and Neugebauer, *Calendars*, p. 86.
[21] I am indebted to Professor Tréheux for permission to use this evidence here.
[22] See the discussion by Dinsmoor, *Athenian Archon List*, p. 191.
[23] See Pritchett and Meritt, *Chronology*, pp. 121-127; B. D. Meritt, *Hesperia*, XXVI, 1957, p. 72; Chryses Pelekides, *B.C.H.*, LXXXI, 1957, pp. 478-484.

Pelekides has argued that one at least of the texts in which the name of Demetrios has been restored as archon belongs with great probability near the middle of the century, and he has dated it (*I.G.*, II², 1027) tentatively in 159/8. But if this year is in fact not available for the archon, the argument, which depends in part on identification of secretaries, may well be incorrect. I have heretofore held that the three inscriptions, *I.G.*, II², 1027, *Chronology*, pp. 121-122, and *Chronology*, p. 125, all belong to one year. Two of them surely belong to the archon Demetrios: *Chronology*, pp. 121-122, and *Chronology*, p. 125; they have the same orator. But *Chronology*, pp. 121-122, names no secretary. So the dating has really depended on making the restoration of the archon's name in *I.G.*, II², 1027, to agree with *Chronology*, p. 125, and on restoring the secretary's name Δη (or Δι) [– – – –]αλ (or αδ) [– – –] of *Chronology*, p. 125, to be the same as Δημήτριος Δημ[– – – –] of *I.G.*, II², 1027. In view of the necessity of moving the archon Demetrios to some year other than 159/8, and in view of the fact further that the only available years are too early (as Pelekides shows) for *I.G.*, II², 1027, this identification must be abandoned. We do not know the name of the archon in *I.G.*, II², 1027; nor do we know the demotic of the secretary. The date is only approximately determinable as near the middle of the second century, and the date of the archon Demetrios must be fixed without reference to it.

Indeed, one must consider the probability that the Demetrios known from *Hesperia*, XXVI, 1957, p. 30 (3), is the same as the Demetrios of *Chronology*, p. 125, now divorced from his supposed association with *I.G.*, II², 1027. There will be the advantage, with only one Demetrios, of not having to explain away the absence of a distinctive qualifying phrase with the second of that name.[24] And the names of the secretaries can

[24] See Pelekides, *B.C.H.*, LXXXI, 1957, p. 484.

be mutually restored, with due regard to the spacing on both stones, as [Διοκ]λῆς Νομ[ίου Δειραδιώτης] in *Hesperia*, XXVI, 1957, p. 30 (3), and as Δι[οκλῆς Νομίου Δειρ]αδ[ιώτης] in *Chronology*, p. 125.

So far as spacing is concerned, in *Chronology*, p. 125, the letters ΔΙ of Δι[οκλῆς] come slightly to the right of the final visible Ο of Δημητρίο[υ] in the line above, and the letters ΑΛ of [Δειρ]αδ[ιώτης] come slightly to the right of ΠΤ in Πτολεμ[αιίδος] in the line above. The space available for name and patronymic and for as much of the demotic as came before the letters ΑΛ is therefore roughly that of the letters [υ ἄρχοντος ἐπὶ τῆς] Πτ of the preceding line. This is precisely correct for Δι[οκλῆς Νομίου Δειρ]αδ[ιώτης] ; the suggestion of Pelekides (*op. cit.*, p. 484) for Δι[οκλῆς Νομίου (?) Π]αλ[ληνεύς] is much too short, and the numerous demotics ending in -άδης are too short for the end of the line. Ἐκαλῆθεν (Ptolemais) finds no available year in the secretary-cycle. Συπαλήττιος (Kekropis) is a possibility with the texts dated in 198/7, the year in which Stamires placed *Hesperia*, XXVI, 1957, p. 30 (3). But it is not permissible to identify Nomios the Sypalettian with the Νόμ[ιος] of *Hesperia*, Suppl. I, p. 117 (61, line 16), or even with one of his family, for whatever the deme there (Aixone ?) it was more numerously represented than one can assume for Sypalettos.

Since we are driven to find an ordinary year, not otherwise occupied, in the early second century, this Demetrios of the Agora texts may be assigned to 190/89, where the cycle demands a secretary from the phyle Leontis, to which the deme Deiradiotai belonged. The character of the writing favors this date.[25]

Something of a point was made by Pelekides (*loc. cit.*) that Demetrios should be dated near in time to Tychandros (160/59)

---

[25] See Pritchett and Meritt, *Chronology*, p. 123.

because two dedications by epheboi (*I.G.*, II², 2981), one of the archonship of Tychandros and one of the archonship of De[– – –], which Pelekides restores as Δη[μητρίου], were cut on the same stone. The inscriptions are on opposite sides of the stone (which is cylindrical), and except that they are dedications by epheboi to Hermes they have no inherent connection with each other. The lettering of De[– – –]'s year is distinctively later and indicates a re-use of the base when the first dedication was obsolete or at least obsolescent. Some little time ought to be assumed between the two dedications, for one group of epheboi could hardly with good grace usurp a monument only within the year set up by their immediate predecessors. I should now restore ἐπὶ Δη[μοστράτου ἄρχοντος] in line 10 and attribute the second dedication to 130/29, or follow my earlier preference and restore (with Kirchner) ἐπὶ Δη[μοχάρους ἄρχοντος] of the year 108/7.[26]

As against the identification of the secretaries of *Chronology*, p. 125, and *Hesperia*, XXVI, 1957, p. 30 (3), there is the prosopographical argument advanced by Stamires (*Hesperia*, *loc. cit.*) that the rare name Νόμιος implies affiliation with the deme Aixone. It is not a compelling argument, and I have preferred, in a sequence of years where so few vacancies exist, not to have two archons named Demetrios if the evidence can be satisfied with only one.

Margaret Thompson has sent to me another of her observations on the Athenian New Style coinage, namely, that the chronological sequence brings an issue with an unusually elaborate cornucopiae of Ptolemaic design to the year 152/1 B.C. This is one of the years that scholars have claimed for the archonship of Lysiades, in whose term it is known that the

---

[26] See Pritchett and Meritt, *Chronology*, p. xxxiv. Indeed, Mitsos, upon examining the stone, believes that the final letters of his name can still be discerned, along with other traces: Δημ[οχ]άρ[ο]υς (letter of November 24, 1959).

Ptolemaia were celebrated at Athens with exceptional splendor.[27] It is perhaps more than a coincidence that the coinage reflects the pomp of the occasion. One objection that has been made to dating Lysiades in 152/1 has been a belief that this year was preëmpted by Phaidrias.[28] But Phaidrias does not belong to an Olympic year. Our evidence for him, derived from names in the athletic victor lists, is that he must be dated between 154/3 and 150/49.[29] There is also preserved a decree (*I.G.*, II², 958) praising the agonothetes of the Theseia in his year. Since the provisions of the decree are that the honors be proclaimed at the Dionysia, the Panathenaia, the Eleusinia, and the Ptolemaia, it has been held that the year of the decree preceded the year in which the Ptolemaia were celebrated. With a known elaborate celebration of the Ptolemaia in 152/1, the decree *I.G.*, II², 958, may well be dated in 153/2, so that the archonship of Phaidrias (of the previous year) thus falls in 154/3. Epainetos, who has been tentatively assigned to 154/3, perhaps belongs in 151/0, and the three archons Aristophantos, Zaleukos, and Mikion, once dated in 151/0, 150/49, and 149/8, can be placed in 150/49, 149/8, and 148/7. None of this can be considered certain, but as the new evidence accumulates the archon-tables of the mid second century lose their lacunae, and there is very little choice about a rearrangement. So far as I am aware, the suggestions here do no violence to the old evidence, and they take account of the new evidence from the coins.

The year 137/6 is the next to be indicated as intercalary by the New Style coinage. This falls within a period of twelve years, from 143/2 to 132/1, where the numismatic evidence is particularly strong. All twelve of the strikings were heavy, and in each case at least eleven, and sometimes twelve or thirteen,

---

[27] Dinsmoor, *Archons*, p. 261. The epigraphical evidence is in *I.G.*, II², 1938.
[28] See, for example, Dinsmoor, *Athenian Archon List*, p. 192.
[29] Dinsmoor, *Athenian Archon List*, p. 192.

months of the year are represented. Between 73 and 156 coins
survive of each issue. The only way to break the sequence,
I am informed, would be to assume a year with no coinage preserved, or to insert here some later issue just before 137/6 or
just before 134/3 (the only places where the issues are not
doubly tied by die sequences) in complete disregard of every
consideration of style. There is, in fact, no other striking even
remotely similar to this remarkably full and homogeneous group.
So, in spite of the fact that I have taken the year 137/6 (archonship of Herakleitos) as ordinary[30] with backward count in the
use of the phrase μετ' εἰκάδας in its one preserved inscription, I
believe that I must reconsider, and accept Dinsmoor's interpretation of the year as intercalary with forward count.[31] The epigraphical evidence is in the text of *I.G.*, II², 974, which should
be read as follows:[32]

*I.G.*, II², 974

*a.* 137/6 *a.* NON-ΣΤΟΙΧ.

Ἐπὶ Ἡρακλείτου [ἄρ]χοντος ἐπὶ τῆς Ἀντιοχίδος ἑ[νδεκάτης πρυτα]
νείας ἧι Διονύσ[ιο]ς Δημητρίου Ἀνακαιεὺς ἐγραμ[μάτευεν· Θαργη]
λιῶνος τρίτει μ[ε]τ' εἰκάδας, ἑβδόμει καὶ εἰκοστ[εῖ τῆς πρυτανείας·]
κτλ.

[Thargelion] 23 = Prytany [XI] 27 = 347th day.

The prytanies regularly had 32 days each, and the last two
months were each of 30 days.

---

[30] B. D. Meritt, *Hesperia*, IV, 1935, p. 560; cf. Pritchett, *Hesperia*, IX, 1940,
p. 132; Pritchett and Meritt, *Chronology*, p. xxxi; Pritchett and Neugebauer,
*Calendars*, p. 86.
[31] W. B. Dinsmoor, *Archons*, p. 417; *Athenian Archon List*, pp. 24, 245-246.
[32] The letters on the preserved portion of the stone, at the ends of lines 1-3,
are so placed that initial epsilon of ἐνδεκάτης falls over the first mu of
ἐγραμμάτευεν, and this mu is half a space to the right of the tau in εἰκοστεῖ. By
count of letters to the right margin, therefore, 13 letters restored in line 1
take up the space of 12 letters restored in line 2, and these take the place of
13½ letters restored in line 3. With the restoration in the *Corpus* (ἑβδόμης for

The numismatic evidence here gives support to forward count with μετ᾽ εἰκάδας. It does, however, add an intercalary year at that point in the calendar where 135/4 is already known as intercalary with backward count μετ᾽ εἰκάδας.[33] And the later numismatic evidence calls for still a third intercalary year in 134/3. But again this falls within the close sequence from 143/2 to 132/1, and there is no more doubt about its attribution than there was for 137/6.

The year of the archonship of Jason (125/4) has been generally accepted as surely ordinary,[34] though I do not now understand the objections I raised in 1933 against the interpretation of it as an intercalary year. If the text from *Hesperia*, II, 1933, p. 163 (9) is restored

Ἐπὶ Ἰάσ[ονος ἄρχοντος ἐπὶ τῆς Κεκρο]πίδος δευτέρας πρυτανεί ας ἧι ᾽Αθην[όδωρος ᾽Αναξικράτους ᾽Ελ]ευ[σίνιο]ς ἐγραμμάτευεν· Μετα γειτνιῶνος [ἑβδόμει ἐπὶ δέκα, τετ]άρτηι κ[αὶ] δεκάτηι τῆς πρυτανείας

the equation is

Metageitnion [17] = Prytany II 14 = 46th day,

and the other known decree of Jason's year (*I.G.*, II², 1003) may be restored

Ἐπὶ Ἰάσονος ἄρχοντος ἐπὶ τῆς ᾽Αντ[ιοχίδος ἕκτης πρυτανείας ἧι
                                                    ᾽Αθηνόδωρος]
᾽Αναξικράτους ᾽Ελευσίνιος ἐγραμ[μάτευεν· Ποσιδεῶνος πέμπτει μετ᾽
                                                    εἰκάδας, τρί]
τει καὶ δεκάτει τῆς πρυτανείας· [— — — — — — — κτλ. — — — — — — — —]

the number of the prytany and Γαμηλιῶνος for the name of the month) the 13½ letters in line 3 are matched by only 11 letters in line 2 and 11 letters in line 1. One would have expected the syllable ΝΕΙ of πρυτανείας at the end of line 1 and the syllable ΛΙ of Γαμηλιῶνος at the end of line 2. Such arguments from spacing cannot be pressed too hard, but the indications here favor the eleventh prytany and the month of Thargelion.

[33] Cf. *Hesperia*, IX, 1940, p. 128 (26).

[34] Meritt, *Hesperia*, II, 1933, pp. 163-165; Pritchett and Meritt, *Chronology*, p. xxxiii; Dinsmoor, *Athenian Archon List*, p. 25; Pritchett and Neugebauer, *Calendars*, p. 86.

[Posideon 26] = Prytany [VI] 1[3] = 173rd day.

I suggest that these restorations be adopted, and that the indication of the numismatic evidence be followed in construing 125/4 as an intercalary year.

Of the two remaining years shown as intercalary by the numismatic evidence, 119/8 is known anyway to have been intercalary on the basis of an inscription published by Werner Peek,[35] and 113/2 may appropriately be restored as intercalary, since it precedes a year that was surely ordinary and follows another surely intercalary year after an interval of two years.

[35] *Kerameikos*, III, pp. 2-3.

# CHAPTER IX

## NEW TEXTS

There is new evidence to present before this discussion of the calendar can be considered down to date. Four newly found inscriptions from the Athenian Agora help in rounding out the documentation and in building up the roster of archons and secretaries that now extends (with epigraphical coverage) from 346/5 down to the beginning of the first century. These new texts are given here.

**1** (See Fig. 3). Part of a stele of Pentelic marble, with the right side and rough-picked back preserved, found among stones removed from the long late Roman wall (O 8) east of the Panathenaic Way in January of 1950.[1] The spring of a crowning moulding is preserved above the first line of the inscription.

Height, 0.24 m.; width, 0.175 m.; thickness, 0.07 m.

Height of letters, 0.005 m.

Inv. No. I 6259.

The surface of the stone is badly worn, and many of the letters can be deciphered only with difficulty.

### OINEIS

*a.* 271/0 *a.*                              nearly ΣΤΟΙΧ. 38-40

['Επὶ Πυθαράτου ἄρχοντος ἐπὶ τῆς Οἰ]νεῖδος [δωδε]

[κάτης πρυτανείας ἧι Ἰσήγορος Ἰ]σοκράτου Κ[εφα]

[λῆθεν ἐγραμμάτευεν· Σκιροφο]ριῶνος ἔνει καὶ [ν]

---

[1] The coördinates of the place of discovery refer to the grid pattern of the Agora as drawn by John Travlos and published by R. E. Wycherley, *The Athenian Agora*, Vol. III: *Testimonia*, Plate II. The grid pattern is also conveniently accessible in *Hesperia*, XXII, 1953, Plate 12.

[ἔαι προτέραι, μιᾶι καὶ τριακο]στεῖ τῆς πρυτανε
5 [ίας· ἐκκλησία· τῶν προέδρων ἐ]πεψήφιζεν Κηφισομ
[— — — — — — — — — — — — —] καὶ συμπρόεδροι· [ἔδοξ]
[εν τῶι δήμωι· — — — — — —]ης εἶπεν· περὶ ὧν ἀπαγ[γ]
[έλλουσιν οἱ πρυτάνεις τῆς] Οἰνεῖδος ὑπὲρ τῶν [θυ]
[σιῶν ὧν ἔθυον τὰ πρὸ τῶν ἐκκλη]σιῶν τοῖς θ[εο]ῖς [οἷς]
10 [πάτριον ἦν· ἀγαθεῖ τύχει δεδό]χθαι τῶι δήμωι τὰ μ
[ἐν ἀγαθὰ δέχεσθαι ἃ φασιν γεγον]έναι ἐν τοῖς ἱερ
[οῖς οἷς ἔθυον ἐφ᾽ ὑγιείαι καὶ σωτη]ρίαι τῆς βουλῆ[ς]
[καὶ τοῦ δήμου καὶ τῶν ἄλλων ὅσοι εὔ]νους [ε]ἰσὶν τ[ῶι]
[δήμωι· ἐπειδὴ δὲ οἱ πρυτάνεις τάς τε] θυσ[ίας] τὰ[ς κα]
15 [θηκούσας ἔθυσαν καλῶς καὶ φιλοτίμω]ς, ἐπεμελ[ήθη]
[σαν δὲ καὶ τῶν ἄλλων ἁπάντων ὧν αὐτοῖς οἵ] τε [νόμοι]
[προσέταττον καὶ τὰ ψηφίσματα τοῦ δήμου — — — — —]

This is one of the series of "prytany" decrees, the chief
interest in which here lies in the calendar equation between
the festival and the conciliar years. The name of the chairman
of the proedroi (otherwise unknown) must have been unusually
long, and the name of the orator unusually short. Restorations
in lines 10-17 have been made with reference to *I.G.*, II², 674,
of 273/2 B.C.[2]

The calendar character of the year 271/0 is known as intercalary (with irregularities) from two texts published by Dinsmoor in 1954,[3] which give also the name of the secretary of this
year, here to be supplied in lines 2-3. The present decree was
passed on the next-to-last day, with the restoration made almost
certain by the near-stoichedon arrangement of the text. The
equation is the same as that attested for the next-to-last day of
303/2 in *I.G.*, II², 495-497.

Hence it is clear that the irregularities within the year, and

[2] For the date, see Dinsmoor, *Hesperia*, XXIII, 1954, p. 314.
[3] *Hesperia*, XXIII, 1954, pp. 284-316. See above, pp. 151-152.

the lengths of months and prytanies, cannot be defined as by Dinsmoor (*op. cit.*, pp. 311-312). The evidence supports the view of Pritchett and Neugebauer (*Calendars*, p. 68) that the prytanies had regularly 32 days each, with observable irregularities to be attributed to the festival calendar. We have suggested above that days added early in Elaphebolion were compensated by omissions late in Elaphebolion. At any rate, this inscription is proof that the correction was not postponed until days had to be dropped from the end of the year.[4]

**2** (See Fig. 4). Fragment of Hymettian marble found on April 19, 1955, among stones collected in the southeastern area of the market square. The inscribed face and rough-picked back are preserved, with the line of a moulding above the inscription. Above the moulding the surface is lost, and below the fourth line of the inscription the surface is badly damaged.

Height, 0.235 m.; width, 0.19 m.; thickness, 0.08 m.
Height of letters, 0.007 m.
Inv. No. I 6731.

a. 271/0 a.                    NON-ΣTOIX. *ca.* 32-34

[θ] ·            ε            [o            ι]
['Επὶ Πυθ]αράτου ἄρχοντος ἐ[πὶ τῆς Οἰνεῖδος]
[δωδεκά]της πρυτανεί[ας· Σκιροφοριῶνος]
[δεκάτ]ει ὑστέραι, τρίτ[ει καὶ εἰκοστεῖ τῆς]
5    [πρυτανείας· — — — — — — — — — — — — — — —]

[Skirophorion] 21 = Prytany [XII] [2]3 = 373rd day.

The calendar equation of the text can be restored with certainty, and offers welcome additional evidence for the character of this year. Inasmuch as Skirophorion ἕνη καὶ νέα προτέρα was

---

[4] I do not agree with the contention that irregularities, wherever occasioned, were usually compensated near the end of the year. See W. K. Pritchett, *B.C.H.*, LXXXI, 1957, p. 282 note 4. See also below, p. 195.

Fɪɢ. 3. Decree of 271/0 ʙ.ᴄ. (Agora Inv. No. I 6259)

Fig. 4. Decree of 271/0 b.c. (Agora Inv. No. I 6731)

FIG. 5. Decree of 181/0 B.C. (Agora Inv. No. I 6765)

Fig. 6. Decree of 177/6 B.C. (Agora Inv. No. I 6166)

equated with Prytany XII 31 (see the preceding text) during the prytany of Oineis, it is clear that Οἰνεῖδος is here to be restored at the end of line 2 and Σκιροφοριῶνος at the end of line 3. The ordinal numeral of the prytany in line 3 is [δωδεκά]της.

This inscription is proof that the irregularities within the year had all been compensated at least earlier than Skirophorion 21.[5] It is also proof that in a hollow month δεκάτη ὑστέρα was not the omitted day, for Skirophorion of this year was planned as a hollow month.[6]

3 (See Fig. 5). Upper part of a stele of Hymettian marble, found on April 30, 1957, near the southwest corner of the Agora (E 17).[7] One small fragment on the left edge was broken off and replaced by mending in antiquity. The text belongs to the well-known series of so-called "prytany" inscriptions.

Height, 0.50 m.; width, 0.48 m.; thickness, 0.13 m.

Height of letters, 0.006 m.

Inv. No. I 6765.

### LEONTIS

a. 181/0 a.                           NON-ΣΤΟΙΧ. 55-63 (ca. 60)

Ἐπὶ Ἱππίου ἄρχοντος ἐπὶ τῆς Λεωντίδος ἕκτης πρυτανείας
  ἧ[ι Θεοδόσιος]
Ξενοφάντου Λαμπτρεὺς ἐγραμμάτευεν· Ποσιδεῶνος ἐμβο-
  λίμου [ἐν]δεκά
τει, ἐνάτει καὶ εἰκοστεῖ τῆς πρυτανείας· βουλὴ ἐμ βουλευτηρίωι σύνκλητος
στρατηγῶν παραγγειλάντων καὶ ἀπὸ βουλῆς ἐκκλησία
  κυρία ἐν τῶι θεάτρωι· vv

---

[5] See above, pp. 151-152, 194.
[6] See the preceding text. See also above, pp. 46-47.
[7] See above, p. 192 note 1.

5  τῶμ προέδρων ἐπεψήφιζεν Ἀμεινίας Ξάνθου Μαραθώνιος
   καὶ συμπρόεδροι· ᵛ
   ἔδοξεν τῶι δήμωι· ᵛ Λακράτης Μέντορος Περιθοίδης εἶπεν·
   ὑπὲρ ὧν ἀπαγ
   γέλλουσιν οἱ πρυτάνεις τῆς Λεωντίδος ὑπὲρ τῶν θυσιῶν ὧν
   ἔθυον τὰ πρὸ τῶν
   ἐκκλησιῶν τῶι τε Ἀπόλλωνι τῶι Προστατηρίωι καὶ τεῖ
   Ἀρτέμιδι τεῖ Βουλαίαι
   καὶ τοῖς ἄλλοις θεοῖς οἷς πάτριον ἦν· ᵛ ἀγαθεῖ τύχει δε-
   δόχθαι τῶι δήμωι τὰ
10  μὲν ἀγαθὰ δέχεσθαι τὰ γεγονότα ἐν τοῖς ἱεροῖς οἷς ἔθυον ἐφ'
   ὑγιείαι καὶ σωτη
   ρίαι τῆς τε βουλῆς καὶ τοῦ δήμου καὶ τῶν συμμάχων, ἐπειδὴ
   δὲ οἱ πρυτάνεις
   τάς τε θυσίας ἔθυσαν ἁπάσας ὅσαι καθῆκον ἐν τεῖ πρυ-
   τανείαι καλῶς καὶ φι
   λοτίμως, ἐπεμελήθησαν δὲ καὶ τῆς συλλογῆς τῆς τε βουλῆς
   καὶ τοῦ δήμου
   καὶ τῶν ἄλλων ἁπάντων ὧν αὐτοῖς προσέταττον οἵ τε νόμοι
   καὶ τὰ ψηφίσμα
15  τα τοῦ δήμου· ἐπαινέσαι τοὺς πρυτάνεις τῆς Λεωντίδος καὶ
   στεφανῶσαι χρυ
   σῶι στεφάνωι κατὰ τὸν νόμον εὐσεβείας ἕνεκεν τῆς πρὸς
   τοὺς θεοὺς καὶ φιλο
   [τιμίας τῆς εἰς τὴν βουλὴν καὶ τὸ]ν δῆμον τὸν Ἀθηναίων·
   ἀναγράψαι δὲ τόδε τὸ
   [ψήφισμα τὸν γραμματέα τὸν κατὰ πρυτα]νείαν ἐν στήλει
   λιθίνει καὶ στῆ
   [σαι – – – – – – – – – – – – – – – – – – – – – – – – – – – –]

The year of Hippias was an intercalary year in the festival
calendar, in which the prytanies each contained 32 days. This
decree was passed either on the 11th or the 12th of the inter-

calated Posideon. If the date is restored as the 11th the equation will be the same as that of a similar date attested in *I.G.*, II², 785, as the 189th day of the year.[8] The equation in *Hesperia*, Suppl. IV, p. 145, should then be

Gamelion 25 = Prytany VIII [8]

restoring the first two lines to read

Ἐπὶ Ἱππίου ἄρχοντ[ος] ἐπὶ τῆς Πανδιονίδος ὁ[γδόης πρυτανείας ἧι Θεοδόσιος Ξενοφάντου Λαμπτρεὺς]
ἐγραμμάτευεν· Γαμηλιῶνος ἕκτει μετ᾽ εἰκά[δας, ὀγδόει τῆς
πρυτανείας· βουλῆς ψηφίσματα· τῶν προέ].

The day was the 232nd day of the year.

The calendar equation in *Hesperia*, Suppl. I, pp. 96-100 (47), is not preserved with enough precision to be significant, but some day in the eleventh prytany fell in Thargelion, and the first two lines may now be restored

[Ἐπὶ Ἱππίου ἄρχοντος ἐπὶ τῆς Ἀττα]λίδος ἐνδεκάτης πρυτ[ανείας]
[ἧι Θεοδόσιος Ξενοφάντου Λαμπτρε]ὺς ἐγραμμάτευεν· Θαργη[λιῶνος].

The calendar equation of *I.G.*, II², 889, represents possibly the 377th day, with the first three lines restored (note the syllabic division):

[Ἐπὶ Ἱππί]ου ἄρχο[ντο]ς ἐ[πὶ τῆς — — — — δος δωδεκάτης πρυτανεί]
[ας ἧι Θε]οδόσιος Ξενοφά[ντου Λαμπτρεὺς ἐγραμμάτευεν· Σκιρο]
[φ]οριῶνος ὀγδόει μετ᾽ εἰκ[άδας, πέμπτει καὶ εἰκοστεῖ τῆς πρυτανεί].

This restoration presupposes a hollow month Skirophorion, in which the omitted day was ἐνάτη μετ᾽ εἰκάδας (see above, p. 53) and in which the count was backward. If the count with μετ᾽ εἰκάδας was forward the day was the 383rd, and the date within the prytany should be restored as μιᾶι καὶ τριακοστεῖ.

---

[8] For the arrangement of the months see *Hesperia*, V, 1936, p. 425. See also Pritchett and Neugebauer, *Calendars*, p. 75.

The calendar equation in *Hesperia*, IX, 1940, p. 355 (49), is not preserved; but the lacunae for restoration in lines 1-2 are very short, and it is quite possible that this text, like the new text published above, is of the sixth prytany, with the reading (note the syllabic division):

[Ἐπὶ Ἱππίου ἄρχοντος ἐπὶ τῆς Λεω]ντίδος <sup>vv</sup>
[ἕκτης πρυτανείας ἧι Θεοδόσιο]ς Ξενοφά[ν]
[τοῦ Λαμπτρεὺς ἐγραμμάτευεν· βο]υλῆς ψη <sup>v</sup>
[φίσματα· – – – – – – – – – – – –] *vacat*
5　[– – – – – – – – – – – – – – – – – –]πε <sup>v</sup>

The new inscription, which belongs on the evidence of its lettering to the early part of the second century, must be assigned either to 193/2 or to 181/0 to satisfy the secretary cycle, which demands a secretary from the phyle Erechtheis (I) to which the deme Lamptrai belongs. If the text is assigned tentatively to 193/2 it will displace Phanarchides, who now holds that year according to the table published in *Hesperia*, XXVI, 1957, p. 94, and also Diodotos, who must follow immediately upon Phanarchides, in whatever years they both belong.[9] The only reasonable displacement of these two archons would be to the years 181/0–180/79, but this new dating is improbable in that it would bring four ordinary years in the festival calendar in succession (see *Hesperia*, *loc. cit.*) while displacing in turn the archons Aphrodisios and Dionysios previously attributed to those years. The displacement of Aphrodisios and Dionysios is no very serious matter—indeed it has already been suggested that Aphrodisios may belong to 170/69[10]—but the necessity for having four ordinary years together is awkward.

On the other hand, if the new text is dated in 181/0 there is not so serious an anomaly in the calendar. Aphrodisios would

---

[9] Cf. Pritchett and Meritt, *Chronology*, p. xxvi.
[10] See *Hesperia*, XXVI, 1957, p. 38 note 28.

still have to move down to 170/69—and I now suggest that this change be made—while Dionysios of *I.G.*, II², 888, had best also be given another berth in order to avoid two intercalary years in succession.[11]

There is no assurance as yet about the year to which this Dionysios belongs. Until more is known about the archons of the early second century it will probably save confusion if he is placed tentatively in 194/3, which is an intercalary year, and where his epithet τοῦ μετὰ – – – – (which must be supplied in line 1 of *I.G.*, II², 888) is justified in order to distinguish him from Dionysios (without the epithet) of 197/6. This arrangement leaves Phanarchides and Diodotos in 193/2 and 192/1 respectively, with Diodotos distinguished by the epithet τοῦ μετὰ Φαναρχίδην.[12] The necessity for making such a distinction implies another archon Diodotos not much earlier in date whom I now assign tentatively to the year 202/1. The table of archons on pp. 235-236 below embodies these changes.

One may note that a date for Hippias either at the beginning or the end of a secretary cycle might have been inferred from the fact that the inscription published in *Hesperia*, Suppl. IV, pp. 144-147, is an inventory, of the kind which normally come at the beginning or ends of cycles.[13] The new text makes clear the attribution that was already indicated in principle by the evidence at hand, fortunately in a more precise form by the preservation of the secretary's demotic. The orator of the decree was Lakrates son of Mentor of Perithoidai (line 6), already known as the orator of the decree concerning the inventory and of the two other decrees of the archonship of Hippias,[14] and as proedros in the archonship of Symmachos in 188/7.[15]

---

[11] He was dated in 180/79 by Pritchett and Meritt, *Chronology*, p. xxvii.

[12] Cf. Pritchett and Meritt, *Chronology*, pp. 114-115.

[13] See Pritchett and Meritt, *Chronology*, p. 58; G. A. Stamires, *Hesperia*, XXVI, 1957, p. 38 note 28.

[14] *I.G.*, II², 889, and *Hesperia*, Suppl. I, pp. 96-100 (47). Cf. *Hesperia*, Suppl. IV, p. 146.      [15] *I.G.*, II², 891; see above, pp. 154-155.

The chairman of the proedroi Ameinias son of Xanthos of Marathon is otherwise unknown. The formula for the convocation of the Council and for the place of meeting of the Council and Ekklesia (lines 3-4) is the same as that of *I.G.*, II², 897, of the year 185/4.

**4** (See Fig. 6). Fragment of Hymettian marble, broken on all sides, found on April 30, 1949, in the long late Roman wall (O 8) east of the Panathenaic Way.[16] The back of the stone is rough, but is perhaps the original surface of the stele.

Height, 0.26 m.; width, 0.134 m.; thickness, 0.104 m.
Height of letters, 0.008 m.
Inv. No. I 6166.

*a*. 177/6 *a*.                    NON-ΣΤΟΙΧ. *ca*. 58-71

['Επὶ – – – ί]ππου ἄρχοντος [ἐπὶ τῆς – – – – – – – – – –
        – – – – – – – πρυτανείας ἧι]
[– – – – ης 'Η]γήτορος Οἰναῖος ἐ[γραμμάτευεν· – – – – –
        ὦνος – – – – – – – –, – – – –]
[– – – – – – – –]ει τῆς πρυτανείας· [ἐκκλησία· τῶν προέ-
        δρων ἐπεψήφιζεν – – – – – – – – –]
[– – – – – – – –]σιος καὶ συμπ[ρόεδροι· ἔδοξεν τῶι
        δήμωι· – – – – – – – εἶπεν· ὑπὲρ ὧν]
5   [ἀπαγγέλλουσι]ν οἱ πρυτάνεις τῆς [– – – – – ὑπὲρ τῶν
        θυσιῶν ὧν ἔθυον τὰ πρὸ τῶν ἐκκλη]
[σιῶν τῶι τε 'Από]λλωνι τῶι Προ[σ]τ[ατηρίωι καὶ τεῖ
        'Αρτέμιδι τεῖ Βουλαίαι καὶ τεῖ]
[Φωσφόρωι καὶ] τοῖς ἄλλοις [θεοῖς οἷς πάτριον ἦν· ἀγαθεῖ
        τύχει δεδόχθαι τῶι δήμωι τὰ μὲν]
[ἀγαθὰ δέχεσθ]αι τὰ γεγο[νότα ἐν τοῖς ἱεροῖς οἷς ἔθυον
        ἐφ' ὑγιείαι καὶ σωτηρίαι τῆς τε βου]
[λῆς καὶ τοῦ δήμου κα]ὶ τῶν σ[υμμάχων – – – – – – – – –
        – – – – – – – – – – – – – – – –]

---

[16] See above, p. 192 note 1.

The number of the letters in the lines shows some variation, in particular being below normal in line 1, which is noticeably more widely spaced. There are two erasures on the preserved surface, apparently where errors have been corrected.

In one of the erasures fewer letters were inscribed than were erased, and in the other there is no perceptible change in the number of the letters. Perhaps the restoration at the end of line 4 should also reflect an error and its correction. With a normal spacing and a normal name for the orator there is no room for the formula of resolution ἔδοξεν τῶι δήμωι; but a mistake of omission could have been corrected, or it may be that the orator's name was unusually short. I have restored the line on the latter assumption.

The text is from the same year as that published in *Hesperia*, XVI, 1947, p. 188 (94), and together they yield the name of the archon [– – – i]ppos and the name of the secretary [– – – –]ης Ἡγήτορος Οἰναῖος. It is best not to define the number of missing letters, because in both inscriptions the position of the left margin is (within limits) uncertain.[17]

---

[17] I have given the archon as [– – – i]ppos in the table below on p. 236, rather than as [. .ª. i]ppos, as it was shown in *Hesperia*, XXVI, 1957, p. 94. An examination of the stone, made in 1958, casts doubt on the reading of a lone tau at the end of line 1 in *Hesperia*, XVI, 1947, p. 188 (94). Since syllabic division was so nearly universal at the ends of lines in this period, the line should be made to end, I think, with the name of the phyle, and the complete ordinal numeral, whatever it was, should be restored at the beginning of line 2.

# CHAPTER X

## The Seasonal Year

There is more to be said about the fifth century. In his article "Calendars of Athens Again,"[1] W. K. Pritchett once more cites the famous logistai inscription, which he and Neugebauer reconsidered in their *Calendars of Athens*. He quotes the dates which we know for correspondences between the conciliar and festival calendars,[2] and which indicate an approximate solar year:

423/2 [Skirophorion] 23 = Prytany X 20

412/1 Skirophorion 14 = Prytany I 1

and then goes on to say, "We could discover nothing in the period before 407 B.C. which would contradict the pattern of a prytany-year of 366 days in which the first six prytanies had 37 days each, the last four 36. Furthermore, there was no evidence that any prytany-year was ever reduced to 365 days, so we avoided referring to it as a solar year."

This logistai inscription is the theme of our fourth chapter, above, where some of the epigraphical and mathematical difficulties of the Pritchett-Neugebauer reconstruction are exposed. Until one can produce a different—and better—interpretation, this inscription is still proof that the two years 424/3 and 423/2, at least, had exactly 365 days each. I should agree, now, that one should perhaps not refer to this conciliar year as a solar year. The term "solar year" has astronomical overtones that were foreign to the political practice of Athens. But the conciliar year of the fifth century may perhaps be defined as a "seasonal"

[1] *B.C.H.*, LXXXI, 1957, p. 292.
[2] *B.C.H.*, LXXXI, 1957, p. 294.

year. The Council began its term about midsummer. This was true in the fifth century as well as in the fourth, third, and second. There was, however, a major distinction about the fifth century. The conciliar year was not tied to the lunar year either at its beginning or at its end. This is now recognized by all students of the calendar, and both in 422 and in 411 B.C. it does not press the evidence too hard, I think, to say that the conciliar year began soon after the solstice.[3]

But if the conciliar year began in midsummer in 422, and was, according to Pritchett and Neugebauer, longer than the solar year by about three days in each quadrennium, then in the 86 years since 508 B.C. the conciliar year will have retarded its inception by about 63 days. Carrying the reckoning back to the institution of the conciliar year of the ten prytanies under Kleisthenes, one finds that this year must have begun (if regularly of 366 days) much nearer to the vernal equinox than to the summer solstice.[4] There is no evidence that the conciliar year wandered thus across the seasons, and indeed the need to assume such vagary would be rather an argument, if one were still needed, against a rigid conciliar year of 366 days.

The fundamental equations for determining the average length of the conciliar year in the fifth century are those just given for the years 422 and 411. These are, indeed, as Pritchett says, "the very cornerstone of fifth century chronology for scholars who have supported a schematic festival calendar."[5]

[3] Pritchett, *Cl. Phil.*, XLII, 1947, p. 243 note 40 (cf. *B.C.H.*, LXXXI, 1957, p. 295): "In the fifth century, however, the date of the beginning of each prytany year cannot be determined; so I see no way at present to correlate the Julian calendar except as a rough approximation."

[4] The date of the institution of the seasonal conciliar year has been much debated. In *The Athenian Calendar*, p. 124 (cf. *Athenian Financial Documents*, p. 153), I advocated the time of Kleisthenes, an opinion which, after some doubts, I still hold. I have not favored Dinsmoor's suggestion of 432 B.C. (cf. *Athenian Financial Documents*, pp. 155-156), and Pritchett and Neugebauer (*Calendars*, pp. 105-106) have convinced me that my argument for a date in mid fifth century was not sound (*Hesperia*, V, 1936, pp. 376-377).

[5] *B.C.H.*, LXXXI, 1957, p. 296.

Yet it is misleading to suggest that acceptance of these two equations implies support of a schematic calendar, festival or otherwise. They are evidence for corresponding dates, nothing more. Pritchett denies their validity. Having dwelt at length on dates in the festival calendar κατὰ θεόν (normal and correct) and κατ' ἄρχοντα (diverging from the date κατὰ θεόν because of intercalation of extra days by the archon), he has come to regard with suspicion all dates in the festival calendar that are not specifically designated as κατὰ θεόν.

But if one looks to the preserved equations of the fourth century, even as outlined by Pritchett and Neugebauer,[6] it is remarkable how many of the festival dates show no irregularity when matched against a hypothetically rigid conciliar year, and still more remarkable how few abnormalities there are (results of tampering) when the prytany calendar of the fourth century is conceded that flexibility which we know to have been valid from other centuries.[7] The known dates may have been dates κατ' ἄρχοντα, but by and large they show no divergence from normal and may be taken also as dates κατὰ θεόν. In any particular instance we are entitled, a priori, to regard a given date as correct; there may even be outside evidence, circumstantial or otherwise, to show that a date should be taken at its face value. If such evidence is at hand, one should not reject it simply because tampering with festival dates by the archon can sometimes be demonstrated elsewhere.

In the case of the two equations in 422 and 411, their validity is assured by the good sense they make in defining the conciliar year. We have already found, mathematically, that the conciliar year was 366 days in 426/5, 368 days in 425/4, and 365 days in 424/3 and 423/2. It was a seasonal year, not tied rigidly to the solstice but (I believe) beginning not far from

---

[6] *Calendars*, pp. 40-41.
[7] See above, p. 131.

it. The first terminal date in the span of eleven years from 422 to 411 is in lines 78-79 of the logistai inscription: [8]

[Skirophorion] 23 = Prytany X 20

and the lower terminal date is in an equation given by Aristotle: [9]

Skirophorion 14 = Prytany I 1.

With the days in the months of both years astronomically correct the beginnings of these conciliar years can be reckoned as separated by 4020 days,[10] an average of $365\frac{5}{11}$ days for each year. This is so close to the average of the solar (and sidereal) year that we are justified not only in assuming that the Athenians intended the conciliar year to be so determined but also that the dates which give the equations are themselves correct.

It would be perverse here to find that these dates were tampered dates. One who posits a rigid conciliar year of 366 days must, of course, assume tampering, for 4020 is not exactly divisible eleven times by 366 days. To save a rigid conciliar year (which we know in fact to have been flexible) one must give up the approximation that these equations yield to a seasonal year, and assume that the known festival dates are inaccurate; perversity is obligatory.

I wish to examine first some of the minor difficulties. In the Aristotelian equation the date of Prytany I 1 is given as the equivalent of Skirophorion 14. This is not a record of something

---

[8] For the text, see Meritt, *Athenian Financial Documents*, p. 141, lines 78-79: ἐπὶ τῆς Λεοντίδος πρυτανείας δεκάτε[ς πρυτανευόσες Σκιροφοριῶνος ὀγ]δόει φθ[ίνοντος εἰ]κοστεῖ τῆς πρυτανείας.

[9] 'Αθ. Πολ., 32, 1: ἔδει δὲ τὴν εἰληχυῖαν τῷ κυάμῳ βουλὴν εἰσιέναι τετράδι ἐπὶ δέκα Σκιροφοριῶνος.

[10] Meritt, *The Athenian Calendar*, p. 85, has 4019 days. The span from June 28, 422, to July 25, 411 (dates of observable first lunar crescents at Babylon, according to Parker and Dubberstein, *Babylonian Chronology*, p. 33), is 4045 days. This is longer than the span of the eleven conciliar years by the first 9 days of Hekatombaion in 422 and the last 16 days of Skirophorion in 411. Hence the length of the eleven conciliar years was $4045 - (9 + 16) = 4020$ days.

that actually happened; it is a prediction of what should have happened, had not the revolution of the Four Hundred intervened. Aristotle (and his source) could predict accurately only if they were dealing with an accurate calendar. To say that the new Council was to enter office on Skirophorion 14 has no meaning if one does not know whether the archon will add or subtract days in the meantime. The unexpired term of the old Council was known. Pritchett holds that the tenth prytany must have had 36 days; I believe that it may well have had 36 days, but with the prytany already well begun I agree that its length was in fact anyway known. Hence the length of Thargelion was known (see above, pp. 36-37) and the number of days down to Skirophorion 14 could be accurately counted. The date was a predictable date κατὰ θεόν.

But if one agrees that no prophecy of corruption is latent in the date of Skirophorion 14 in 411, because of its citation as a fixed point in the forecast of the conciliar year, it may perhaps be thought that the date Skirophorion 23 in 422 is incorrect. If all the intervening conciliar years had 366 days apiece, it would indeed have to be taken as six days in error, the actual date κατὰ θεόν being, by count of days back from 411, Skirophorion 17. We should then have, in effect, a combination of dates: Σκιροφοριῶνος ὀγδόει φθίνοντος κατ᾽ ἄρχοντα κατὰ θεὸν δὲ ἑβδόμει ἐπὶ δέκα. But here the archon date is later by six days than the lunar date, and such a maladjustment never occurs. Archon dates varied from lunar dates, if at all, by being earlier, rather than later, than reality.[11]

---

[11] Leonardo Taran, of the Princeton Graduate School, has questioned this tenet. While it is true that there is no formal proof and that all the epigraphical examples so far known run the other way, one should not (he claims) deny the sovereign Demos the power to omit as well as to add if it so thought fit. Taran cites Aristotle, ᾽Αθ. Πολ., 40, 1, as proof that Archinos did subtract days from the calendar in 403/2, thus preventing many Athenians from leaving Athens when democracy was restored in the archonship of Eukleides. The simple device of Archinos was to eliminate from the calendar the final days of registration. If these days were specified in terms of the festival year, then

Again, if Skirophorion 23 was the same by both the archon and the moon, and the archon decided to tamper with the calendar by adding extra days, he might add, for example, six days so that the sixth later day would still be called Skirophorion 23 κατ' ἄρχοντα, but the true date would have been advanced to Skirophorion 29 κατὰ θεόν. This suits the Pritchett-Neugebauer scheme even less well, and we have no example of over-compensation when the correction of an irregular addition was effected by omission of later days. It has sometimes been inferred that days added early in the year were compensated only late in the year.[12] There is no evidence for this. Where irregularly intercalated days were added the correction was made as soon as conveniently possible. It is difficult to believe that month after month would pass, with the count κατ' ἄρχοντα at noticeable odds with the lunar cycle at every easy point of verification (especially new moon and full moon) without any attempt at correction, especially when one attaches importance, as Pritchett and Neugebauer do, to the regulation of festival months by the observed crescent and especially since the temporary occasion for the intercalation of extra days would be long since passed and no longer valid.

We are left then with this additional dilemma when we assume rigid conciliar years in the fifth century: we must either reject the testimony of Aristotle in his account of the revolution of the Four Hundred, or assume that he was making a prediction based upon a calendar which was, *ex hypothesi*, unpredictable, ignoring the predictable date κατὰ θεόν.

the dates after the deletion were later κατ' ἄρχοντα than the true dates κατὰ θεόν. There is no suggestion that Archinos changed the law about registration so as to deny the privilege while the days remained untouched: as Aristotle says, ὑφεῖλε τὰς ὑπολοίπους ἡμέρας τῆς ἀπογραφῆς. A device similar to that of Archinos would also solve very neatly the obdurate calendar problems of 318/7 (see above, p. 56 note 29) and of 171/0 (see above, p. 161 note 59).

For a truly extraordinary example of adding and subtracting in the Athenian calendar, at the whim of the Demos, see Plutarch's account (*Demetrius*, XXVI) of the initiation of Demetrios into the Mysteries.

[12] See now W. K. Pritchett, *B.C.H.*, LXXXI, 1957, p. 282 note 4, and p. 297.

But this is not all. In the record so far as preserved, in all centuries, fifth, fourth, third, and second, there is no example of extra days added irregularly to the festival calendar or of compensating subtraction in the month of Skirophorion, which ended the year. This is true even at those times when tampering with the calendar was most rife. There were doubtless many reasons for adding days to the calendar, but, as Pritchett has noted, they were for the most part connected with the postponement of festivals, possibly in time of inclement weather, and disturbances were especially apt to occur in winter. Our own study has shown that most of the additions in which three or more days were involved came at the time of the City Dionysia and to a lesser degree at the time of the Panathenaia.[18] The fact of interest now, in its bearing upon the dates in Skirophorion in the fifth century, is that, with all the irregularities in the festival calendar and with all the tampering which the archons permitted themselves down through the centuries, there is no single instance—not one—in which a date in Skirophorion is not valid. So far as we have evidence, every known date in Skirophorion was not only κατ' ἄρχοντα but κατὰ θεόν as well.

I have suggested above (pp. 172-175) a solution of the calendar problems of 222/1 and 221/0 which, under special circumstances, will have left the dates within the month valid but have changed the name of the month. And we do not know how omissions in the year 166/5 compensated for the early addition of days before Anthesterion (pp. 175, 183-184), but otherwise the record is clear, and I urge that even those who accept the applicability of archons' tampering with least reservation (in principle) have no evidence that suggests irregularity in the dates of Skirophorion 23 in 423/2 and of Skirophorion 14 in 412/1.[14]

---

[18] See above, pp. 161-166.

[14] I have translated the equation Σκιροφοριῶνος ὀγδόη φθίνοντος = Prytany X 20 of 422 B.C. into one which shows the commencement of the conciliar year 17 days later: Hekatombaion 10 = Prytany I 1 (*The Athenian Calendar*, p. 84).

Beginning with 422/1, I have followed the lapse of three years to 419/8, and shown in a table in *Athenian Financial Documents*, p. 176,[15] that in 419 we probably have the equation

Hekatombaion 16 = Prytany I 1.

This equation is in agreement with that derived from Antiphon's speech Περὶ τοῦ χορευτοῦ

Metageitnion 21 = Prytany I 36

and both are cited for an "unknown year" by Pritchett and Neugebauer in their *Calendars of Athens*, p. 108.[16] Now Pritchett has again turned to these calendar equations,[17] using them as an *exemplum horribile* of my method "of using a schematic festival calendar to obtain dates for historical events." What I have done is to go on the assumption that the dates given are honest dates, and I have simply used normal years to reach the agreement with Antiphon. I do not doubt that some chaotic scheme could give a similar equation in some other year, but it is illogical to assume chaos, which may yield any result we fancy, when normal order meets the normal requirements:

(1) that one of the three years 422/1, 421/0, and 420/19 be intercalary. This is almost inevitable, since 424/3 and 423/2 were both ordinary. These two ordinary years are de-

---

Pritchett questions this equation because (a) he believes that one cannot determine whether Skirophorion was full or hollow and (b) he thinks that the archon, in tampering with the calendar, may have omitted days from the end of Skirophorion or added days at the beginning of Hekatombaion (*B.C.H.*, LXXXI, 1957, p. 297). It makes no difference (a) whether Skirophorion was full or hollow: the date ὀγδόῃ φθίνοντος shows that the month still had seven more days to run (see above, p. 45). There is no evidence (b) that the Athenians ever added days before Hekatombaion 10; the record is just as clear as it is against omitting days in Skirophorion.

[15] See also *The Athenian Calendar*, p. 118.

[16] It is a misprint that the first equation is given by Pritchett and Neugebauer as Hekatombaion (I) 1 = Prytany I 1 (for the date see also A. E. Raubitschek in *Hesperia*, XXIII, 1954, p. 70).

[17] *B.C.H.*, LXXXI, 1957, pp. 297-298.

cisively proved by the necessity for dating the first payment in 424/3 in the logistai inscription 705 days before the end of the quadrennium and yet before the Panathenaic festival. Had either 424/3 or 423/2 been intercalary the payment would have fallen after the festival.

(2) that the conciliar years be taken as seasonal years. This is proved by the logistai inscription and by the 11-year span from 422 to 411 B.C.

As a matter of fact, the three prytany years from 422 to 419 are all assumed in my table to have been of 366 days. Had they been taken as years of 365 days, some of the later years after 419 would have had to be of 366 days in order to meet Aristotle's equation in 411. But I do not now claim, and never have claimed, that the table or its determinations are rigidly exact in the intervals between fixed points.

Pritchett writes as follows: "At the close of the chapter in *The Athenian Calendar* which we have been discussing, the author sets up a table of synchronisms in terms of the Julian calendar, his schematic festival or civil calendar, and the prytany calendar. Using this table, he concludes his study in the final two pages by assigning a date to Antiphon's speech 'On the Choreutes.' This speech contains a reference to days in terms of both the prytany calendar and the festival calendar, and the majority of scholars have deduced therefrom the equation Prytany I 36 = Metageitnion (II) 21. Using his table of synchronized dates, Meritt can conclude: 'In the table given above on p. 118 we find that 419/8 is the only year in which this chronological correspondence existed, and in this year I propose that the speech must be dated. For literary history the determination is valuable as giving a fixed point for the study of the development of rhetorical style in Athens during the latter part of the Fifth Century.'"

The table is, in fact, a kind of summary and epitome of the evidence. It was clear that any normal progression of festival and conciliar years, either before or after 422/1, would not come even close to Antiphon's requirements except in 419/8. So it may be said that we have placed the pieces of the puzzle where they fit, and the equation Prytany I 1 = Hekatombaion 16 was put into the table as a refinement of normal calendar development because here alone all the evidence could be taken into account.

Pritchett's comment on the date of Antiphon's speech continues: "Later, Dover made this date of 419/8 B.C. a basis for a study of the chronology of Antiphon's speeches [the reference is to *Cl. Quart.*, XLIV, 1950, pp. 44-60]. He defended Meritt's date against the 'threat' of Pritchett and Neugebauer, as he termed it, in a complete *peritrope*: he accepts Meritt's date, then evokes Meritt's schematic calendar containing the synchronism Pryt. I 1 = Hekatombaion (I) 10 (422/1 B.C.), and Meritt's arrangement of intercalary years to prove that Meritt's date was correct."

I should not have stated the case quite in this way. What Dover uses in his argument is simply the equation. We know that it is a good equation because there is no evidence anywhere for any but good equations in the festival calendar at Athens during Skirophorion or the early days of Hekatombaion. The spectre of tampered dates, and the fear of them, which hovers over the argument here like a kind of miasma, should be dispelled.

From the equation in 422/1 B.C. Dover then proceeds logically through the three succeeding years in the festival and prytany calendars, and concludes, as I have done, that the year 419 is correct, and uniquely so, for Antiphon's speech. In this sense it may be said that he accepts my date for the speech; but there is no *peritrope*.

Pritchett continues: "But the present state of our studies does not permit us to equate dates of the festival and prytany calendars of 419 B.C. and neighboring years, and the date of Antiphon IV [*sic*: a misprint for VI] would seem to be still unsolved." It is indeed unsolved if one assumes that the known dates in the festival calendar are incorrect: guilty, that is, until proved innocent. In view of the fact that such dates in late Skirophorion and early Hekatombaion were never wrong, I find it astonishing that one has come to deny the *prima facie* evidence of these texts.

In studying the correspondences between the conciliar and the festival years in the fifth century Pritchett and Neugebauer have stopped with the year 408/7, but they have expressed confidence that the year 407/6 as well can be made to agree with their system.[18] The epigraphical evidence is in *I.G.*, I², 304B, for which I have elsewhere given photographs, drawings, and a text.[19] Now Pritchett has again examined the stone in Paris, and has come to doubt some of my readings.[20] I can only say at this writing that the stone is indeed difficult to decipher, but that the readings as I have given them were made without prejudice and that they are not at all what I had expected to find. Instead of showing that the separate conciliar year was continued to the end of the century, the readings indicate rather that the festival and conciliar years were coterminous in 407/6. The significant calendar equations are those of the months Metageitnion, Boedromion, and Mounichion; they belong apparently to an ordinary year of 354 days, which I have diagrammed as follows: [21]

[18] *Calendars*, p. 94 note 5.
[19] Meritt, *Athenian Financial Documents*, pp. 120-122, with Plates VII-XI.
[20] *B.C.H.*, LXXXI, 1957, p. 296 note 1.
[21] *Athenian Financial Documents*, p. 124.

407/6   Prytany I  1 = Hekatombaion 1

| | | | |
|---|---|---|---|
| Prytany I | 37 | Hekatombaion | 29 |
| Prytany II | 36 | Metageitnion | 30 |
| Prytany III | (35) | Boedromion | (29) |
| Prytany IV | (35) | Pyanepsion | (30) |
| | | Maimakterion | (29) |
| Prytany V | (34) | Posideon | (30) |
| Prytany VI | (34) | Gamelion | (29) |
| Prytany VII | (34) | Anthesterion | (30) |
| Prytany VIII | 35 | Elaphebolion | (29) |
| | | Mounichion | 30 |
| Prytany IX | (37) | Thargelion | (29) |
| Prytany X | (37) | Skirophorion | (30) |
| | 354 | | 354 |

406/5   Prytany I  1 = Hekatombaion 1.

This scheme for the year differs from the Aristotelian norm in having one certain prytany of 37 days, as well as two other such long prytanies (restored) and three short prytanies of 34 days each (restored). I feel less hesitation now about the one extra day in the first prytany, after having found that the first prytany of 195/4 had one more than a normal length of days,[22] and that the same may have been true of 258/7 and of 176/5.[23] We have discovered that in the fourth, third, and second centuries Aristotle's rule of sequence of prytanies proved itself flexible. Can we believe that the Athenians would have bound themselves not ever to vary the length of a prytany? In his sympathetic review of Pritchett and Neugebauer's book Jean Pouilloux noted the freedom with which ordinary and intercalary years were allowed to succeed each other with reference

[22] See above, p. 140, and p. 144 note 27.
[23] See above, pp. 140-142, 144-145.

to no established cycle, and then asked: "Cette attitude ne témoigne-t-elle pas de l'approximation que les Athéniens ne surent jamais exclure de leur chronologie? Le cherchaient-ils même et ne serait-ce pas une erreur d'esprits modernes et scientifiques que de vouloir introduire dans le calendrier attique une pensée rigoureuse et systématique? " [24]

If one is willing to assume moderate tampering with the festival calendar, he may posit the addition of 3 days before Mounichion 3 and their elimination by way of compensation after Mounichion 22. The diagram of the year (almost a normal year) will then appears as follows:

407/6   Prytany I 1 = Hekatombaion 1

| | | | |
|---|---|---|---|
| Prytany I | 37 | Hekatombaion | 29 |
| Prytany II | 36 | Metageitnion | 30 |
| Prytany III | (35) | Boedromion | (29) |
| | | Pyanepsion | (30) |
| Prytany IV | (35) | Maimakterion | (29) |
| Prytany V | (35) | Posideon | (30) |
| Prytany VI | (35) | Gamelion | (29) |
| Prytany VII | (35) | Anthesterion | (30) |
| Prytany VIII | 35 | Elaphebolion | (29) |
| | | Mounichion | 30 ± 3 |
| Prytany IX | (35) | Thargelion | (29) |
| Prytany X | (36) | Skirophorion | (30) |
| | ——— | | ——— |
| | 354 | | 354 |

406/5 Prytany I 1 = Hekatombaion 1.

The year 408/7 shows a sixth prytany of 37 days and an eighth prytany of 36 days. This argues that the separate con-

[24] *B.C.H.*, LXXIII, 1949, p. 498. See also the judicious comment of A. G. Woodhead, *The Study of Greek Inscriptions* (1959), pp. 115-118.

ciliar year was still in existence at that time, and in 1928 I so interpreted it.[25] Perhaps so; and perhaps my argument is not valid that the separate conciliar year ended with Hekatombaion 1 of 409 b.c.[26] The strongest evidence for denying the existence of a separate conciliar year begins in 407/6. It may be that this was the first, rather than the third, year of the new dispensation.

But for the years earlier than 411 b.c. the seasonal dates over a considerable period can be equated approximately with Julian dates.

Meton's first cycle began with the summer solstice of 432 b.c. on Skirophorion 13. Since this was the beginning of his astronomical calendar, and since it is hardly conceivable that Meton deliberately began to name his months at variance from those of the festival calendar, it may be taken as certain that in 432 b.c. the new year (Hekatombaion 1) began with the new moon following this date. According to Parker and Dubberstein,[27] the crescent moon was first visible on July 17, and this is the date on which I now believe the archonship of Pythodoros (432/1) to have begun.[28]

Various arguments that have been used to show the sequence of ordinary and intercalary years after 432 are no longer sound.

<hr />

[25] *The Athenian Calendar*, pp. 99-100. The salaries paid by the prytany were scaled accurately to the number of days in the prytany. Such accuracy was not always the rule: in the late fourth century the architect's salary for 13 months, at the rate of 2 drachmai a day, was reckoned as 780 drachmai (*I.G.*, II², 1673, line 60). This would imply 13 consecutive months of 30 days each, which we recognize as a bookkeeping device and not as a reflection of the calendar.

[26] *Athenian Financial Documents*, pp. 104-109, 176.

[27] *Babylonian Chronology*, p. 32.

[28] The Julian date of Hekatombaion 1 in 432 b.c. was shown in *Athenian Financial Documents*, p. 176, as July 15. But the table there was scaled to astronomical conjunctions rather than to first visibility. Meton's plan for an astronomers' calendar was based on lunar months and his cycles were timed to begin and end with conjunctions (see above, p. 16). The difference between conjunction and first visibility in 432 was comparable to that at the end of the seventeenth cycle in 109. See Dinsmoor, *Archons*, p. 410.

In the first place, my own argument for making 432/1 itself ordinary, based upon Cavaignac's interpretation of Thucydides (II, 24) is not valid,[29] for we now know the decree mentioned by Thucydides not to be the same as that summarized in the Strasbourg papyrus.[30] Nor can we use the astronomical evidence of eclipses of the sun (Thuc. II, 28) and moon (scholiast on Aristophanes, *Clouds*, 584) for the festival calendar of 431 and of 425 B.C., because the first of these is no longer tied to an Athenian archonship and the second was almost surely given in terms of the schematic astronomical calendar which the Athenians did not necessarily follow in their own festival year.

The conciliar year was regulated roughly as a seasonal year and depended for its epochal date on the summer solstice. This means that the new year which began on Hekatombaion 1 of 422 B.C. can be fixed by the equation of the logistai inscription. If the eighth day from the end of Skirophorion was the same as the twentieth day of the tenth prytany, then the new moon which began the following festival year was that near the end of June. According to Parker and Dubberstein,[31] the new crescent was visible on June 28, and this is the date on which I believe the festival year 422/1 to have begun.[32] The reckonings implicit in the logistai inscription show that 424/3 and 423/2 were both ordinary years,[33] and I now give the initial dates (after Parker and Dubberstein) as follows:

424/3   Hekatombaion 1   July 19, 424
423/2   Hekatombaion 1   July  9, 423.

From 422 down to 411 the sequence of ordinary and intercalary years may be given as in *Athenian Financial Documents*,

[29] *Athenian Financial Documents*, pp. 174-175.
[30] Wade-Gery and Meritt, *Hesperia*, XXVI, 1957, pp. 181-188.
[31] *Babylonian Chronology*, p. 33.
[32] The Julian date of Hekatombaion 1 in 422/1 was shown in *Athenian Financial Documents* as June 26.
[33] See above, pp. 209-210.

p. 176, but from there to the end of the century uncertainty about the succession of the years increases. It has been our belief, as expressed above, that the year 407/6 was ordinary.

Earlier than 432/1, it may be shown that the year 433/2 was intercalary. From *I.G.*, I², 295, it is known that the Treasurers of Athena, who changed office at the Panathenaia, were different on Prytany I 13 and Prytany I 37 (assuming this to be the last day). The conciliar year, being a seasonal year, began soon after the solstice. We shall be sufficiently near the truth, for present purposes, if we fix the date tentatively, let us say, at July 4. Prytany I 13, therefore, will have fallen on July 16 and Prytany I 37 on August 9. Hekatombaion 28, the date of the Panathenaia, will have fallen between these two Julian dates, and the Hekatombaion 1 which preceded will have fallen, consequently, between July 19 and July 13. The available new moon is that which first became visible, according to Parker and Dubberstein,[34] on June 28. Since we thus have the equation of Hekatombaion 1 of 433/2 with the Julian date June 28, it follows that the year 433/2 was itself intercalary with 384 days (June 28, 433 B.C. to July 16, 432 B.C.).[35]

The correspondences between the conciliar year, the festival year, and the Julian calendar may now be shown in a revised table based upon that in *Athenian Financial Documents*, p. 176.

[34] *Babylonian Chronology*, p. 32.
[35] Meritt, *The Athenian Calendar*, pp. 88-89.

| | | FESTIVAL YEAR | | | CONCILIAR YEAR | | |
|---|---|---|---|---|---|---|---|
| | | Attic inter-calation | Date of Hekatom-baion 1 | Number of days | Date of Prytany I 1 | Number of days | Julian inter-calation |
| | 433/2 | I | June 28 | 384 | Hek. 7 = July 4 | 365 | O |
| First Metonic Cycle | 432/1 | (O) | July 17 | 355 | Skir. 17/18 = July 4 | 365 | O |
| | 431/0 | (O) | July 7 | 354 | Skir. 27/28 = July 4 | 365 | O |
| | 430/29 | (I) | June 26 | 384 | Hek. 9 = July 4 | 366 | I |
| | 429/8 | (O) | July 14 | 354 | Skir. 20/21 = July 4 | 365 | O |
| | 428/7 | (I) | July 3 | 384 | Hek. 2 = July 4 | 365 | O |
| | 427/6 | (O) | July 22 | 354 | Skir. 12/13 = July 4 | 365 | O |
| | 426/5 | (I) | July 11 | 384 | Skir. 23/24 = July 4 | 366 | I |
| | 425/4 | O [36] | July 29 | 355 | Skir. 5/6 = July 4 | 368 | O |
| | 424/3 | O | July 19 | 355 | Skir. 18/19 = July 7 | 365 | O |
| | 423/2 | O | July 9 | 354 | Skir. 28/29 = July 7 | 365 | O |
| | 422/1 | I | June 28 | 384 | Hek. 10 = July 7 | 366 | I |
| | 421/0 | O | July 16 | 354 | Skir. 21/22 = July 7 | 366 | I |
| | 420/19 | O | July 5 | 354 | Hek. 4 = July 8 | 366 | O |
| | 419/8 | I | June 24 | 384 | Hek. 16 = July 9 | 365 | O |
| | 418/7 | O | July 13 [37] | 354 | Skir. 26/27 = July 9 | 366 | I |
| | 417/6 | I | July 1 | 384 | Hek. 9 = July 9 | 365 | O |
| | 416/5 | O | July 20 | 355 | Skir. 19/20 = July 9 | 365 | O |
| | 415/4 | I | July 10 | 384 | Skir. 29/30 = July 9 | 365 | O |
| | 414/3 | I | July 29 | 384 | Skir. 10/11 = July 9 | 366 | I |
| Second Metonic Cycle | 413/2 | O | Aug. 16 | 354 | Tharg.22/23= July 9 | 365 | O |
| | 412/1 | O | Aug. 5 | 354 | Skir. 3/4 = July 9 | (365) | O |
| | 411/0 | O [38] | July 25 | 354 | ⟨Skir. 14⟩ = July 9 | (– –) | O |
| | 410/09 | O | July 14 | 355 | (Skir. 15) = June 29 | (370) | I |
| | 409/8 | (I) | July 3 | 384 | (?) Hek. 1 = July 3 | (384) | O |
| | 408/7 | (O) | July 22 | 354 | (?) Hek. 1 = July 22 | (354) | O |
| | 407/6 | O | July 11 | 354 | Hek. 1 = July 11 | 354 | O |
| | 406/5 | (I) | June 30 [39] | 384 | Hek. 1 = June 30 | 384 | I |
| | 405/4 | (O) | July 18 [40] | 355 | Hek. 1 = July 18 | 355 | O |
| | 404/3 | (–) | | | Hek. 1 = July 8 | (– –) | O |

[36] I do not press the evidence of *I.G.*, I², 63, with its mention of simple Posideon, in favor of 425/4 as an ordinary year (see Dinsmoor, *Archons*, p. 333 note 6; and for the text of *I.G.*, I², 63, see Meritt, Wade-Gery, McGregor, *Athenian Tribute Lists*, II [1949], p. 41 [A9 line 19]). Nor is the scholiast's note on the *Clouds* of Aristophanes (line 584) that an eclipse of the moon took place in Boedromion of 425/4 of value for the Athenian calendar (see Meritt, *The Athenian Calendar*, pp. 89-90); this eclipse, of Julian date October 9, 425 B.C., probably belongs within the astronomical reckoning of the Metonic cycle. But the statement of Philochoros that Kleon attacked the proposal to make peace with Sparta ἐπ' Εὐθύνου ἄρχοντος brings the first of Hekatombaion in 425 more than usually late in the year. The festival year 425/4 must have been ordinary if the reference of Philochoros concerned the Spartan overtures of peace (but see A. W. Gomme, *A Historical Commentary on Thucydides*,

The differences between this table and that which I published earlier are principally the use of observable new moon (approximately) rather than time of conjunction for beginning each festival year and the abandonment of any attempt to fix the order of full and hollow months within individual years.[41]

Though I do not believe that Meton's astronomical cycle was authoritative for the succession of ordinary and intercalary years in the festival calendar, it should be noted that the years of his sixth cycle are in fact all definitely known in the calendar of Athens (337/6–319/8).[42] They happen to show a quite regular sequence, and this absence of any irregularity is a guarantee that the first day of Hekatombaion fell near the observable new moon of July 16 in 337 B.C.

We have found that the astronomical data of Ptolemy's *Almagest* have no bearing on the festival calendar of Athens,[43] but there is one event of importance which is associated with a lunar eclipse in 168 B.C. This is the battle of Pydna, in which Aemilius Paulus and the Romans with their allies defeated Perseus of Macedonia. The Athenians honored by decree of the Council and Demos a certain Kalliphanes of Phyle, an Athenian citizen, who had campaigned with the Romans and who had been present at the battle. It was he who brought the news of the

III [1956], p. 488). For the sequence of time, see Wade-Gery and Meritt, *A.J.P.*, LVII, 1936, pp. 377-394, especially pp. 383-384.

[37] This is one day later than the observable new moon as shown by Parker and Dubberstein, *Babylonian Chronology*, p. 33.

[38] Pritchett and Neugebauer, *Calendars*, p. 108, note that in my *Athenian Calendar*, p. 96, I rejected Aristotle's implication ('Aθ. Πολ., 33) that this year was ordinary. They do not note my recantation in *Athenian Financial Documents*, pp. 104-106.

[39] This is one day earlier than the observable new moon as shown by Parker and Dubberstein, *Babylonian Chronolgy*. p. 33.

[40] *Ibid.*

[41] This means leaving out of account the "ideal" calendar of months taken from Dinsmoor's study which I reproduced in *Athenian Financial Documents*, pp. 177-179.

[42] See above, pp. 132-134.

[43] See above, pp. 18-33.

victory to Athens.[44] The decree was voted in the Ekklesia on
the last day of the festival year 169/8, which I have heretofore
taken as ending with the last day of the waning moon early in
July. Our present arrangement of the years of the fourteenth
Metonic cycle (185/4–167/6; see below, p. 236) shows that
a double intercalation in 171/0 and 170/69 had retarded the
festival calendar. The first day of the new year 168/7, therefore, began with the new moon that was first visible on August 7.
So the decree was passed on August 6. If the account in Livy
is to be believed that the battle was fought on the day following
a lunar eclipse (that of the evening of June 21, 168 B.C.),[45] this
was approximately six weeks after the date of the battle. But
Livy also dates the battle after the summer solstice;[46] so the
matter must, I think, remain in doubt. With the later date, the
interval between the battle and the honors voted to Kalliphanes
in Athens can be reduced.[47]

[44] *Hesperia*, V, 1936, pp. 429-430 (cf. III, 1934, pp. 18-21).
[45] Livy, XLIV, 37.
[46] Livy, XLIV, 36.
[47] As in *Hesperia*, III, 1934, p. 20, I again refer to a discussion by W. S.
Ferguson on the problem of the date of the battle (*Athenian Tribal Cycles*,
p. 11 note 1).

# CHAPTER XI

## THE SEQUENCE OF YEARS

In addition to the table given in the preceding chapter for years in the fifth century and in addition to the table given above (pp. 132-134) for determining Julian dates during the sixth Metonic cycle, a skeleton compendium of reference may now be given to show the Athenian archons (as dated) and the types of years in the festival calendar from the mid fourth to the early first century.[1] There are a number of new combinations, but in particular some special explanation is required for a change I have suggested from the accepted date of the archon Arrheneides in the mid third century and for an interpretation I have given of the secretary-cycle in the mid second century.

### THE ARCHONSHIP OF ARRHENEIDES

The traditional date accepted for Arrheneides is 262/1 B.C. In his year the philosopher Zeno died, and Kleanthes became head of the Stoic School, a position which he held until his death in the archonship of Jason. The best short account of the chronology, thus based on the lives of the philosophers, is in Dinsmoor's *Archons of Athens* (pp. 46-48),[2] though one must

[1] The calendar character of individual years is shown at the left margin by the letters O (ordinary) and I (intercalary). Where I have considered the character of the year certain, or strongly supported by the evidence, I have added an asterisk. Reference should be made throughout to Pritchett and Meritt, *Chronology of Hellenistic Athens* (1941) for further details; general reference should also be made to Pritchett and Neugebauer, *Calendars of Athens* (1947) and to the partial tables in *Hesperia*, XXIII, 1954, pp. 312-316, and XXVI, 1957, pp. 94-97. Specific mention of these sources as well as of other studies is sometimes made below under individual years. I have not included references to the latest tables given by Eugenio Manni in *Athenaeum*, XXXVII, 1959, pp. 245-257.

[2] But Dinsmoor's references to *Ind. Stoic.* III and IV should read rather to Philodemos, Περὶ τῶν Στωικῶν.

221

still weigh the evidence and be warned that many readings offered from the papyri from Herculaneum are in fact extremely doubtful. Nor is it possible longer to hold that Philokrates was archon four years before Arrheneides. Dinsmoor and I both date him now in 276/5.

The vital clue to the date of Arrheneides is the statement by Philodemos, Περὶ τῶν Στωικῶν (col. IV), that the span of time from the archonship of Klearchos (301/0) to the archonship of Arrheneides was 39 years and three months. Beloch (*Gr. Gesch.*, IV, 2, p. 56) quotes, with a slight variation, the text given by Crönert:

ἀπὸ
10  Κλεάρχου γὰρ ἐπ' ['Αρρ]ενε[ί–
δην, ἐφ' οὗ συμ[φωνοῦ]σι κατα[τε–
τελευτηκένα[ι] Ζήνωνα, ἔτη
ἐστὶν ἐννέα κα[ὶ] τριάκο[ντα
καὶ μῆνες τρεῖς.

Crönert has a somewhat different version in *Kolotes und Menedemos* (1906), p. 138: ἀπὸ Κλεάρχου γὰρ ἐπ' 'Αρρενείδην, ἐφ' οὗ σημ[ειοῦσι κ]αὶ τετελευτηκέναι Ζήνωνα, ἔτη ἐστὶν ἐννέα κα[ὶ] τριάκο[ντα] καὶ μῆνες τρεῖς (τρ. sicher gelesen).

I doubt that the text is so sound as it appears. Felix Jacoby (*Apollodors Chronik*, 1902, p. 363) gives nothing after ἐφ' οὗ in line 11, and has only τηκεν......ήνων in line 12, with a general comment: " leider ist die kolumne so verstümmelt, dass sich nicht viel mit ihr anfangen lässt – – –." Von Arnim (*S. V. F.*, Vol. I, 1905 [1938], p. 13) has in lines 11-14:

δην, ἐφ' οὗ Σκει(ροφοριῶν)ι (κ)ατα(τε–
τελε(ύτη)κεν κ... Ζ)ήνων απ
ἐστὶν ἐ(ν)ν(έα καὶ) (τ)ριάκ(οντα.
καὶ μὴν ἐ...........νοσε.

Here the reading Σκει(ροφοριῶν)ι is surely wrong, but the

significant observation is that the left edge of the column is legible (and reliable) while the center and the right edge are not readily legible and are in consequence unreliable. The letters ἤνων απ in line 12 have become Ζήνωνα ἔτη as though there were no question about them. The reading καὶ μῆνες τρεῖς in line 14 is probably correct, though I doubt the last three letters of τρεῖς. But whether the span was 39 years (line 13), or perhaps 41, or indeed something else again, seems still subject to legitimate doubt, with the corollary that a date for Arrheneides in 262/1 should not be held as a fixed point. A reading of 40 (ἔ[τ]η [τεσσα]ράκο[ντα]) or of 41 (ἐν [καὶ τεσσα]ράκο[ντα]), for example, would permit a date for Arrheneides in 260/59. Indeed, if the papyrus is only slightly in error, this date may be possible even with the reading of 39.[3]

Of more influence on Arrheneides is the virtual certainty that the archon Peithidemos belongs in 265/4,[4] at the beginning of a secretary-cycle.[5] The epigraphical texts of his year have uninscribed stone where the name of the secretary should have been cut,[6] and since this is the year of the commencement of the Chremonidean War it looks as if Dinsmoor was right, in spite of his own erroneous date for Peithidemos, in beginning the new cycle in 265/4 with Erechtheis (III) rather than with Macedonian Antigonis (I).[7] The absence of the secretary's name implies that the man originally chosen for the year (from Antigonis) did not serve. We now continue the secretary-cycles

---

[3] For other wide variants in the reading of this text see F. Jacoby, *Frag. gr. Hist.*, II D, pp. 736-737, who refers to the passage as " sehr unsicher und immer wieder anders gelesen."

[4] *Hesperia*, XXVI, 1957, p. 97.

[5] These cycles, either singly or compounded, marked the span of the great public inventories (Pritchett and Meritt, *Chronology*, pp. 22-46). This is true here in the third century of the inventory of the priests of Asklepios (*I.G.*, II², 1534 B), which covered the dedications from the archonship of Peithidemos to the archonship of Diomedon, and of the inventory of offerings on the acropolis (Pollux, X, 126) drawn up in the archonship of Alkibiades.

[6] *I.G.*, II², 687, line 3; *Hesperia*, V, 1936, p. 419, no. 14.

[7] *Hesperia*, XXIII, 1954, p. 314

224 The Sequence of Years

past 255/4, when Alkibiades was archon at the beginning of a new cycle with Antigonis (I). Athens was again on good terms, officially, with Macedonia.[8] And the cycles continue uninterruptedly until the known break between Hieron and Diomedon, in whose year they begin again with Erechtheis (III).

One version of the *Index Stoicorum* gives the time of the primacy of Kleanthes as 32 years. This is the span from Arrheneides to Jason. When Arrheneides was held to 262/1 the span of 32 years dated Jason (and the death of Kleanthes)[9] in 230/29. But again the readings of the papyrus are uncertain. Traversa, in his edition of 1952, gives the years as [τ]ριάκ[ον]τα καὶ [δύ]ọ,[10] and says of the [δύ]ο "ipse vidi." Beloch (*Gr. Gesch.*, IV, 2, p. 56) has given Crönert's reading as τριάκ[ο]ντα καὶ [ἔ]ν, which Traversa does not dispute, though Beloch's comment on the short numeral is "zu δύο und ἕξ passen die erhaltenen Spuren nicht, so dass nur ἕν möglich bleibt." And he adds that Crönert later verified this reading.[11] August Mayer punctuated after τριάκοντα, and read the complete numeral as 30.[12]

It is difficult to make out from the several publications here just how much has (supposedly) been seen. Theodor Gomperz has left a vivid description of his attempts to learn what the reading should be.[13] He wrote a letter of enquiry to Comparetti asking what the "legerissimi indizi" might be that led him in his publication to think of ὀκτώ. Comparetti's reply stressed that these "indizi" were so vague that he could not describe them; that he had at Naples three times examined the passage with the purpose of securing sure evidence for the age of Zeno,

---

[8] See the references to the renewal of friendship in *I.G.*, II², 477, of the following year (cf. *S. E. G.*, III, 89, and *Hesperia*, VII, 1938, pp. 141-142).

[9] *Index Stoicorum Herculanensis*, edidit Augustus Traversa (Istituto di Filologia Classica, Genova, 1952), col. XXVIII.

[10] *Op. cit.*, col. XXIX.

[11] See W. Crönert, *Kolotes und Menedemos* (1906), p. 192, *s. v.* Kleanthes.

[12] *Philologus*, LXXI, 1912, pp. 236-237. Cf. Dinsmoor, *Archons*, p. 48 note 5.

[13] *Rh. Mus.*, XXXIV, 1879, p. 155.

but "toujours sans résultat." Gomperz then himself, in the winter of 1876/7, examined the passage, accompanied by Dr. Corazza, custodian of the papyrological collection of the Museo Nazionale. They were agreed on one definite conclusion: that there was in this numeral a horizontal stroke, quite certain and clear, that could be the bottom stroke of delta. For Gomperz this was proof that his conjecture for the restoration δύο was correct. Corazza thought the stroke might belong to tau, or pi, or upsilon, all of which in this papyrus have a kind of flourish in a horizontal stroke at the bottom.

The views of Crönert and Gomperz, both of whom have studied the original, are irreconcilable, and Corazza's judgment could lead to rejecting the numeral altogether. In spite of Traversa's indication of a new paragraph after [δύ]ο, the new argument may have begun after [τ]ριάκ[ο]ντα, with some such phrase as καίτ[οι Δ]ιο[ν]ύσ[ιος] τοίν[υν ὁ − − − − −], or καὶ τ[ὸν Δ]ιο[ν]ύσ[ιον] τοίν[υν τὸν − − − − −]. And, even on the evidence of the papyrus, there may have been only thirty years from Arrheneides to Jason.

So the archon Jason, perhaps, belongs in 231/0 or 230/29. The same fragment of papyrus avers that Kleanthes was born when Aristophanes was archon (331/0). He is supposed to have died at the age of 99 (or 100),[14] but the uncertainty will perhaps permit a tentative dating of Jason in 231/0, or even in 230/29. I do not regard the measurements of the intervals of time in the lives of the philosophers as having the same claim to accuracy, in our sense of the word, as do the definitions of the limits in terms of Athenian archonships.

If Arrheneides is correctly dated in 260/59, on the evidence of continuity in the secretary-cycles, the end of the Chremonidean War belongs in 261/0, in the archonship of Antipatros, his

[14] See Beloch, *Gr. Gesch.*, IV, 2, pp. 56-57.

immediate predecessor.[15] The war thus lasted four years, from 265 to 261. The earlier dating of it, down to the fall of Athens, has been for the four years from 267 to 263. King Areus of Sparta died during the Chremonidean War in a second attempt to force the isthmus of Corinth. Ferguson claims that his death cannot be later than 264 B.C.;[16] the dating of Peithidemos in 265/4 shows that it cannot have been earlier. If 264 was in fact the date of Areus's death, events in the first year of the war moved more rapidly than Ferguson supposed. But the siege of Athens was protracted, and, because of distractions and defections in the camp of Antigonos, not at first very effective.[17]

A choregic dedication on Delos (*I.G.*, XI, 2, no. 114) mentions the blessings of peace achieved in the archonship of Tharsynon (261 B.C.): [ἐπ᾽ ἄρ]χοντος Θαρσύνοντος τοῦ Χ[οι]-ρύλου ὑγίεια εἰρήνη πλοῦτος ἐγένετο.[18] If this refers to the end of the Chremonidean War—and it is difficult to see how else the peace can be interpreted—the capitulation of Athens and the end of the war are datable within the same year. There is this added indication that Antipatros at Athens belongs in 261/0.

If Arrheneides does not belong in 260/59, the alternative is to date Polystratos and Kleomachos in 261/0 and 260/59 and to move Antipatros and Arrheneides back to 263/2 and 262/1. In this case the secretary-cycle will have commenced with Antigonis (1) in the archonship of Peithidemos (265/4), and a break of two years must be assumed somewhere between 260/59 and 256/5. The events of the Chremonidean War before the fall of Athens must then be compressed within a period of at most a little more than two years, or parts of three campaigning seasons (265/4–263/2).

---

[15] F. Jacoby, *Frag. gr. Hist.*, II BD (Apollodoros 244, frag. 44).

[16] *Hellenistic Athens*, p. 177 note 1.

[17] Ferguson, *Hellenistic Athens*, pp. 178-182.

[18] I am indebted to Peter Fraser for the reference to this text.

## THE SECRETARY-CYCLE IN 145 B.C.

The five Athenian phylai created after the time of Kleisthenes were named after powerful friends and benefactors of Athens: (1) Antigonis in 307 B.C. for Antigonos Monophthalmos, (2) Demetrias in 307 B.C. for Antigonos's son Demetrios Poliorketes, (3) Ptolemais in 223 B.C. for Ptolemy III, (4) Attalis in 200 B.C. for Attalos I, and (5) Hadrianis in A.D. 124/5 for the emperor Hadrian.[19]

One expects, naturally, that the eponyms, if given the privileges of Athenian citizenship, will have been members of the phylai that were named in their honor,[20] and that this membership will have been shared by those in their families who were also Athenian citizens. Hadrian is known to have been accepted into the deme of Besa and the phyle of Hadrianis.[21] Ptolemy V Epiphanes, grandson of Ptolemy III, was a member of the phyle Ptolemais.[22] There is no evidence about Antigonos and Demetrios, though almost every conceivable honor was voted to them by the Athenian demos; they may well have been citizens within their own phylai. Only with Attalis is there a discordant note, for Attalos II, son of the eponym, was of the deme Sypalettos,[23] which belonged not to Attalis but to Kekropis.

The evidence that Sypalettos belonged, in fact, to Kekropis is indeed overwhelming. Passing over any references earlier than 200 B.C., one finds the attribution to Kekropis unanimously attested by the inscriptions from 123/2 (*I.G.*, II², 1006, line 123) down to the end of the second century after Christ (*I.G.*, II²,

---

[19] See W. K. Pritchett, *The Five Attic Tribes after Kleisthenes* (Diss., The Johns Hopkins University, Baltimore, 1943).

[20] Cf. Pritchett, *The Five Attic Tribes after Kleisthenes*, p. 36 note 13; B. D. Meritt, *Hesperia*, XVII, 1948, p. 29.

[21] *I.G.*, II², 1832; *I.G.*, II², 1764; cf. P. Graindor, *Athènes sous Hadrien*, p. 14 note 1.

[22] *I.G.*, II², 2314, lines 41-42: [βα]σιλεὺς Πτολεμαῖος βασιλέως Πτολεμαίου [Πτολ]εμαιΐδος φυλῆς.

[23] *I.G.*, II², 3781.

1782, line 44).[24] The fact that the deme Sypalettos does not appear in the complete roster of prytaneis from the phyle Attalis in 173/2 B.C. (archonship of Alexis) is further proof that at that early date the affiliation was rather with Kekropis.[25] As Stamires remarked, in his publication of the register, it is well-nigh impossible to hold that even part of the deme Sypalettos at that time belonged to Attalis.[26]

Yet the tantalizing fact remains that unless Sypalettos belonged to Attalis in 146/5 there must have been a break in the secretary-cycle between 146/5 and 145/4;[27] indeed there must have been two breaks, for the irregularity in 146/5 is evidence for an earlier break at some time after 155/4,[28] or perhaps after 153/2.[29] The archonship of Epikrates (146/5), following immediately upon that of Archon (147/6),[30] is definitely fixed by the list of Delian gymnasiarchs (*Inscr. Délos*, 2589), for the gymnasiarch Gorgias son of Asklepiades of Ionidai, twentieth in the list after 166/5, was in office when Archon was archon (*Inscr. Délos*, 1952). From these hard facts there is no escape. A Sypalettian was secretary in 146/5, ending one cycle of secretaries (however disposed), and the following cycle began normally in 145/4 with a secretary from Lamptrai (Erechtheis).

If Sypalettos was too small a deme to be divided, and if in fact it was not divided, the only way in which it can be attributed to Attalis in 146/5 is to assume that quite as a *tour de force* the Athenian demos transferred Sypalettos *in toto* from Kekropis

---

[24] Cf. W. B. Dinsmoor, *Athenian Archon List*, p. 177 note 39.

[25] G. A. Stamires, *Hesperia*, XXVI, 1957, p. 47.

[26] Knowledge of this inscription led me in 1957 to give up the idea of a divided deme, and prematurely, I think, to exclude Attalis from consideration (*Hesperia*, XXVI, 1957, p. 96) even in 146/5.

[27] The date 146/5 was written by mistake as 147/6 in *Hesperia*, XVII, 1948, p. 29.

[28] See the record of the secretary-cycle in Pritchett and Meritt, *Chronology*, p. xxx.

[29] See the record of the secretary-cycle in *Hesperia*, XXVI, 1957, pp. 95-96.

[30] *Inscr. Délos*, 1505, and commentary.

to Attalis at some time after 173/2 and restored it again to Kekropis at some time before 123/2. Within this period of fifty years there is, so far as I know, no evidence for the phyle to which Sypalettos belonged, except the *a priori* assumption that it should have belonged to Attalis because Attalos was a demesman of it, and except the normal requirement of the secretary-cycle which leads one to expect a deme from the phyle Attalis in 146/5 B.C.

Lest this idea of a deliberate transfer of a deme from one phyle to another seem too bizarre, let us recall that whole groups of demes were so transferred when new phylai were created. In 307 B.C. large demes had been transferred in their entirety to Antigonis and Demetrias, or had been divided so that part only (without change of name) went to the new phylai; and small demes were transferred *in toto*. Similar transfers took place in 223 to Ptolemais and in 200 to Attalis. So there is nothing new in the idea of transferring a deme from one phyle to another. It was a political manoeuvre which the Athenians understood, and which they employed when the political motivation was sufficiently compelling. If there was any motivation of sufficient strength in the mid second century, there is, therefore, no reason to deny that the Athenians could, if they wished, have shifted the small deme Sypalettos from one phyle to another. Indeed, it would have been a much simpler thing to do than to create a new phyle and to transfer many demes.

My suggestion now is that such motivation did exist, and that it was the desire of the Athenian demos to show honor to Attalos II when he was a student of philosophy at Athens. Apparently it humored the prince, when he was given citizenship, to belong to the same deme with his kinsman Ariarathes of Kappadokia (a Sypalettian) and yet at the same time to the phyle named in honor of his father. Ariarathes was not only a fellow-student with Attalos under Karneades (*I.G.*, II², 3781) but also a relative. His father, Ariarathes IV, had betrothed his daughter, Stratonike, sister of Ariarathes V, to Eumenes II of Pergamon about 188 B.C. (Livy, XXXVIII, 39, 3).[31] This Ariarathes V became king of Kappadokia in 162 B.C. (*P.A.*, 1608), while Attalos II became king of Pergamon in 159 B.C. But before their accessions they had, as Sypalettians, together dedicated a statue of Karneades in the Agora at Athens (*I.G.*, II², 3781).

Granted that Ariarathes had been made a citizen of Athens in the deme of Sypalettos and in the phyle of Kekropis at some time earlier in the second century, and that Attalos also was made a citizen, at a date later than 172 but earlier than 162, with his choice of deme and phyle, the problems of the secretary-cycle and of the logical affiliation of his citizenship would both be solved if Sypalettos was, for a time, out of deference to him, given in fact to the phyle of Attalis.[32] The deme was then re-assigned to Kekropis when Attalos was no longer alive and Attalids no longer members of it.[33] For the secretary-cycle the important consideration is the possible affiliation of Sypalettos with Attalis in 146/5 B.C.

---

[31] After the death of Eumenes she became the wife of Attalos II himself.

[32] The Athenians did not know, when Ariarathes was made a citizen in Sypalettos, that Attalos would have such close ties in friendship, and in family life, and in the Academy, that he too would want to be a Sypalettian. If Ariarathes had been inducted into a deme of Attalis in the first place (any deme), I hold that there would have been no need to transfer Sypalettos later.

[33] Philetairos, brother of Attalos II, was an Athenian citizen (*P.A.*, 14254), but the grant was presumably made after 175/4 (*I.G.*, II², 905) when he was praised but named only as a brother of Eumenes and not as an Athenian. If he too was a Sypalettian who belonged to Attalis, the grant of his citizenship must have been made after 173/2. Ordinary Athenians named their children after both Attalos and Ariarathes, but this implies no blood relationship. See, for example, Ἀριαράθης Ἀττάλου, a pythaist of 128/7 (*Fouilles de Delphes*, III, 2 [12 III 9]), and Ἀριαράθης Πολεμαίου Συπαλήττιος, treasurer of Kekropis in 95/4 (*Hesperia*, XVII, 1948, p. 26 [12 line 46]). Also, the mint-magistrate Ariarathes of the second century was too late to be identified with Ariarathes V (*P.A.*, 1608) of Kappadokia (communication from Margaret Thompson), and, except for the name, probably had no connection with the royal house.

## THE ATHENIAN CALENDAR
### 346/5–81/0

| Type | Year | Archon | Deme of Secretary | | Reference |
|------|------|--------|------|------|-----------|
| O* | 346/5 | Archias | VII | Phlya | *I.G.*, II², 215, 218 (above, pp. 72-73) |
| | 345/4 | Euboulos | VIII | Oion | *I.G.*, II², 219, 220 |
| | 344/3 | Lykiskos | 9 | | |
| | 343/2 | Pythodotos | X | Aigilia | *I.G.*, II², 223, 224, 225 |
| | 342/1 | Sosigenes | I | Euonymon | *I.G.*, II², 227 |
| I | 341/0 | Nikomachos | II | Araphen | above, pp. 10, 73 |
| | 340/39 | Theophrastos | III | Kytheros | *I.G.*, II², 231, 233 |
| | 339/8 | Lysimachides | IV | Cholleidai | *Hesperia*, VII, 1938, pp. 291-292 |
| ? | 338/7 | Chairondes | 5 | | above, pp. 73, 76 |
| O* | 337/6 | Phrynichos | VI | Acharnai | above, pp. 76-78, 132 |
| I* | 336/5 | Pythodelos | 7 | | above, pp. 10-15, 78-79, 132 |
| O* | 335/4 | Euainetos | VIII | Acherdous | above, pp. 79-82, 133 |
| O* | 334/3 | Ktesikles | IX | Phaleron | above, pp. 82-83, 133 |
| I* | 333/2 | Nikokrates | X | Pallene | above, pp. 48-51, 83-85, 133 |
| O* | 332/1 | Niketes | I | Anagyrous | above, pp. 85-88, 133 |
| O* | 331/0 | Aristophanes | II | Myrrhinoutta | above, pp. 88-91, 133, 225 |
| I* | 330/29 | Aristophon | III | Paiania | above, pp. 91-94, 133 |
| O* | 329/8 | Kephisophon | IV | Eupyridai | above, pp. 94-95, 133 |
| I* | 328/7 | Euthykritos | V | Hagnous | above, pp. 95-97, 133 |
| O* | 327/6 | Hegemon | VI | Acharnai | above, pp. 98-101, 133 |
| O* | 326/5 | Chremes | 7 | | above, pp. 6-8, 89, 101-102, 133 |
| I* | 325/4 | Antikles | VIII | Eleusis | above, pp. 102-104, 133 |
| O* | 324/3 | Hegesias | IX | Rhamnous | above, pp. 104-106, 133 |
| O* | 323/2 | Kephisodoros | X | Alopeke | above, pp. 106-110, 133 |
| I* | 322/1 | Philokles | I | Kephisia | above, pp. 110-112, 133 |

### Deme of Anagrapheus

| | | | | | |
|------|------|--------|------|------|-----------|
| O | 321/0 | Archippos | | Oion | above, pp. 112-113, 134 |
| I* | 320/19 | Neaichmos | | Lamptrai | above, pp. 113-120, 134 |
| O* | 319/8 | Apollodoros | | Anakaia | above, pp. 121-125, 134 |

### Deme of Secretary

| | | | | | |
|------|------|--------|------|------|-----------|
| O* | 318/7 | Archippos | II | Kollytos | above, pp. 55-58, 125-127, 134 |
| I | 317/6 | Demogenes | | | |
| O | 316/5 | Demokleides | | | |
| O | 315/4 | Praxiboulos | | | |
| I* | 314/3 | Nikodoros | | | above, pp. 128-129 |
| | 313/2 | Theophrastos | | | *I.G.*, II², 451 |
| | 312/1 | Polemon | | | |
| | 311/0 | Simonides | | | |

| Type | Year | Archon | | Deme of Secretary | Reference |
|------|------|--------|---|-------------------|-----------|
| O* | 310/09 | Hieromnemon | | | above, pp. 129-130 |
| | 309/8 | Demetrios of Phaleron | | | |
| | 308/7 | Kairimos | | | |
| I* | 307/6 | Anaxikrates | II | Diomeia | above, pp. 176-178 |
| O* | 306/5 | Koroibos | XI | Rhamnous | above, pp. 138-139 |
| O* | 305/4 | Euxenippos | XII | Alopeke | Pritchett and Neugebauer, Calendars, p. 79 |
| O* | 304/3 | Pherekles | I | Gargettos | ibidem |
| I* | 303/2 | Leostratos | III | Phegous | above, pp. 150-151 |
| O* | 302/1 | Nikokles | IV | Plotheia | above, pp. 52, 57-58; Pritchett and Neugebauer, Calendars, pp. 80, 83-84 |
| I* | 301/0 | Klearchos | V | Probalinthos | Pritchett and Neugebauer, Calendars, p. 69 |
| O | 300/299 | Hegemachos | 6 | | |
| O* | 299/8 | Euktemon | VII | Kephale | Pritchett and Neugebauer, Calendars, p. 80 |
| O* | 298/7 | Mnesidemos | VIII | Phyle | ibidem |
| I | 297/6 | Antiphates | 9 | | Pritchett and Meritt, Chronology, p. xvi. |
| O | 296/5 | Nikias | X | Azenia | ibidem; above, pp. 178-179 |
| I | 295/4 | Nikostratos | XI | Phaleron | above, pp. 28-33 |

Deme of Anagrapheus

| | | | | | |
|---|---|---|---|---|---|
| O | 294/3 | Olympiodoros | | Phyle | above, pp. 26-27; Pritchett and Meritt, Chronology, p. xvi |
| O* | 293/2 | Olympiodoros | | Rhamnous | Pritchett and Meritt, Chronology, p. xvii |
| I | 292/1 | Philippos | | | Pritchett and Meritt, Chronology, p. xvii; Meritt, Hesperia, XXVI, 1957, p. 54 |

Deme of Secretary

| | | | | | |
|---|---|---|---|---|---|
| | 291/0 | Charinos | XI | Trikorynthos | Hesperia, XXVI, 1957, pp. 53-54 |
| | 290/89 | Telokles | 12 | | ibidem |
| O* | 289/8 | Aristonymos | I | Aithalidai | |
| O | 288/7 | Kimon | II | Thorai | I.G., II², 697 [34] |
| I | 287/6 | Xenophon | 3 | | |

[34] Sterling Dow reports and A. G. Woodhead confirms from an examination of the stone that the right margin lies farther to the right by one letter in I.G., II², 697. The restoration can be made with [ἄρχων Κίμων· ἐπὶ τῆς Ἱππο-θωντίδος δωδεκ]άτη[s | πρυτανείας – – –] and with the month Σκιροφοριῶνος. For the formula with the archon see Hesperia, IV, 1935, p. 562, no. 40.

| Type | Year | Archon | Deme of Secretary | | Reference |
|------|------|--------|------|------|-----------|
| O* | 286/5 | Diokles | IV | Halai | |
| O* | 285/4 | Diotimos | V | Paiania | |
| I | 284/3 | Isaios | 6 | | |
| O* | 283/2 | Euthios | VII | Cholargos | |
| I* | 282/1 | Nikias | VIII | Acharnai | |
| O* | 281/0 | Ourias | IX | Aixone | above, p. 136 note 4 |
| I* | 280/79 | Gorgias | X | Eleusis | *I.G.*, II², 670A |
| O | 279/8 | Anaxikrates ³⁵ | 11 | | |
| O | 278/7 | Demokles | 12 | | |
| O* | 277/6 | Sosistratos | 1 | | |
| I* | 276/5 | Philokrates | II | Melite | above, p. 222 |
| O* | 275/4 | Olbios | III | Euonymon | |
| O | 274/3 | Euboulos | 4 | | |
| I* | 273/2 | Glaukippos | V | Myrrhinous | |
| O* | 272/1 | Lysitheides | VI | Sounion | *I.G.*, II², 689, 704, 816; *Hesperia*, XXVI, 1957, pp. 54-57 |
| I* | 271/0 | Pytharatos | VII | Kephale | above, pp. 151-152, 192-195; *Hesperia*, XXIII, 1954, pp. 288-289, 299-300 |
| O* | 270/69 | Philippides | 8 | | |
| O | 269/8 | Philinos | 9 | | *I.G.*, II², 1304 *b*; Agora Inv. No. I 5592 (to be published soon in *Hesperia*) |
| I | 268/7 | Diogeiton | X | Keiriadai | *I.G.*, II², 772 |
| O* | 267/6 | Menekles | XI | Trikorynthos | see Pritchett and Neugebauer, *Calendars*, p. 84, for the calendar |
| O* | 266/5 | Nikias Otryneus | XII | Alopeke | |
| O* | 265/4 | Peithidemos | 3 | | *Hesperia*, XXVI, 1957, p. 97; above, pp. 221-226 |
| I | 264/3 | Diognetos | 4 | | *ibidem* |
| I | 263/2 | Polystratos | 5 | | above, pp. 221-226 |
| O* | 262/1 | Kleomachos | VI | Kettos | *Hesperia*, XXVI, 1957, p. 97; Pouilloux, *Rhamnounte*, p. 127, suggests a date *ca.* 250 B.C.; above, pp. 221-226 |
| O | 261/0 | Antipatros | 7 | | above, pp. 221-226 |
| O* | 260/59 | Arrheneides | 8 | | above, pp. 137, 221-226 |
| I | 259/8 | Antiphon | 9 | | |
| O | 258/7 | Thymochares | 10 | | above, pp. 140-142 |
| O | 257/6 | Lykeas (?) | 11 | | |

---

[35] *I.G.*, II², 670B, has been claimed for Anaxikrates by Pritchett and Meritt, *Chronology*, p. xviii. The demotic of the secretary's name is very doubtful, and I have not used it as evidence in this table; if it was, in fact, Εἰτεαῖος, then *I.G.*, II², 670B, belongs to the archon Sosistratos of 277/6. There must also be a change in the restoration in line 2 of the decree from the year of Anaxikrates published in *Hesperia*, XVII, 1948, p. 1, no. 1.

| Type | Year | Archon | Deme | of Secretary | Reference |
|------|------|--------|------|--------------|-----------|
| I* | 256/5 | Euboulos | XII | Alopeke | |
| O | 255/4 | Alkibiades | 1 | | Pritchett and Meritt, *Chronology*, p. 97 |
| I | 254/3 | Philostratos | II | Potamos | Pouilloux, *Rhamnounte*, p. 122 |
| O* | 253/2 | [...]bios | 3 | | |
| I* | 252/1 | Kallimedes | IV | Plotheia | above, pp. 145-146; Pouilloux, *Rhamnounte*, p. 184, suggests the date 246/5 |
| O* | 251/0 | Antimachos | V | Myrrhinous | Pouilloux, *Rhamnounte*, p. 122 |
| O* | 250/49 | Thersilochos | VI | Phrearrhoi | Pouilloux, *Rhamnounte*, p. 127 |
| O* | 249/8 | Polyeuktos | VII | Kephale | |
| I* | 248/7 | Hieron | VIII | Oe | above, p. 224 |
| | | | | | |
| O* | 247/6 | Diomedon | III | Anagyrous | above, pp. 137-138, 223 note 5, 224 |
| I | 246/5 | Philoneos | 4 | Hyporeia (?) | |
| O* | 245/4 | Theophemos | 5 | | above, pp. 142-143 |
| O | 244/3 | Kydenor | VI | Eupyridai | above, pp. 146-148 |
| I | 243/2 | Eurykleides | 7 | | |
| O | 242/1 | Phanomachos | 8 | | |
| I* | 241/0 | Lysiades | 9 | | |
| O* | 240/39 | Athenodoros | X | Hamaxanteia | |
| I* | 239/8 | Lysias | XI | Aphidnai | |
| O | 238/7 | Phanostratos | 12 | | Pouilloux, *Rhamnounte*, p. 122 |
| O | 237/6 | Kimon | 1 | | |
| I* | 236/5 | Ekphantos | II | Hippotomadai | Pouilloux, *Rhamnounte*, pp. 129-132 |
| O* | 235/4 | Lysanias | III | Euonymon | |
| O | 234/3 | Pheidostratos | IV | Erchia | Pouilloux, *Rhamnounte*, p. 122 |
| | 233/2 | | 5 | | |
| | 232/1 | | 6 | | *I.G.*, II², 766 + *Hesperia*, XVII, 1948, pp. 5-7 (cf. *Hesperia*, VII, 1938, pp. 114-115) must be assigned to 244/3 |
| | 231/0 | Jason | 7 | | above, pp. 221-226 |
| | 230/29 | | 8 | | |
| O* | 229/8 | Heliodoros | IX | Athmonon | |
| I* | 228/7 | Leochares | X | Oion | |
| O* | 227/6 | Theophilos | XI | Aphidnai | above, p. 143 |
| I* | 226/5 | Ergochares | XII | Alopeke | above, pp. 152-154 |
| O | 225/4 | Niketes | 1 | | |
| O | 224/3 | Antiphilos | 2 | | |
| I* | 223/2 | | III | Kedoi | Pritchett and Neugebauer, *Calendars*, p. 90 |

| Type | Year | Archon | Deme of Secretary | | Reference |
|------|------|--------|------|------|-----------|
| O* | 222/1 | Archelaos | IV | Ankyle | above, pp. 172-175, Pritchett and Neugebauer, *Calendars*, pp. 91-92 |
| I* | 221/0 | Thrasyphon | V | Paiania | above, pp. 173-175 |
| O | 220/19 | Menekrates | 6 | | |
| O* | 219/8 | Chairephon | VII | Kydantidai | above, pp. 168-169 |
| O* | 218/7 | Kalli[– – –] | VIII | Kephale | above, pp. 169-170 |
| I | 217/6 | | 9 | | |
| O | 216/5 | Hagnias | 10 | | |
| I* | 215/4 | Diokles | XI | Keiriadai | |
| O* | 214/3 | Euphiletos | XII | Rhamnous | above, p. 168 |
| O | 213/2 | Herakleitos | 13 | | |
| I | 212/1 | Euandros | 1 | | |
| O | 211/0 | Aischron | 2 | | |
| O | 210/09 | | 3 | | Philinos is now dated in 269/8 |
| I* | 209/8 | [– – ca. 11 – – –] | IV | Erchia | |
| O* | 208/7 | Ankylos | 5 | | above, pp. 170-172 |
| O | 207/6 | | 6 | | |
| I | 206/5 | Kallistratos | 7 | | |
| O | 205/4 | Pantiades | 8 | | |
| I* | 204/3 | Apollodoros | IX | Oe | above, p. 179; *Hesperia*, XVI, 1947, pp. 190-191 |
| O* | 203/2 | Proxenides | X | Aixone | above, pp. 167-168 |
| O | 202/1 | Diodotos (?) | | | above, p. 199 |
| O | 201/0 | Isokrates | V | Aigilia | *Hesperia*, XXVI, 1957, pp. 64, 94; cf. Pritchett and Neugebauer, *Calendars*, p. 84 note 20 |
| I | 200/199 | Nikophon | 6 | | *Hesperia*, XXVI, 1957, pp. 64, 94 |
| O | 199/8 | [. . .i]ppos | VII | Kothokidai (?) | *Hesperia*, XXVI, 1957, pp. 62-63, 94 |
| O | 198/7 | | 8 | | above, p. 186 |
| O | 197/6 | Dionysios | 9 | | *Hesperia*, XXVI, 1957, pp. 64, 94; above, p. 199 |
| I* | 196/5 | Charikles | X | Rhamnous | below, p. 236 note 36 |
| O | 195/4 | [– 9-12 – –] | XI | Semachidai | above, pp. 139-140; *Hesperia*, XXVI, 1957, pp. 69, 94 |
| I* | 194/3 | Dionysios [after – – –] | 12 | | above, pp. 198-199 |
| O* | 193/2 | Phanarchides | I | Lamptrai | above, pp. 198-199 |
| O* | 192/1 | Diodotos after Phanarchides | II | Halai (?) | above, p. 199; *Hesperia*, XXVI, 1957, p. 94 |
| I | 191/0 | | 3 | | |
| O* | 190/89 | Demetrios | IV | Deiradiotai | above, pp. 135 note 2, 184-187 |

| Type | Year | Archon | Deme of Secretary | | Reference |
|------|------|--------|-----|-----|-----------|
| I* | 189/8 | Euthykritos | V | Kydantidai | above, pp. 149-150; *Hesperia*, XXVI, 1957, pp. 63-66 |
| O* | 188/7 | Symmachos | VI | Thorikos | above, pp. 154-158, 199 |
| O* | 187/6 | Theoxenos | VII | Perithoidai | above, p. 138 |
| I* | 186/5 | Zopyros | VIII | Aixone | |
| O | 185/4 | Eupolemos | IX | Hamaxanteia | |
| I* | 184/3 | | X | Rhamnous | above, pp. 181-182 |
| O | 183/2 | Hermogenes | XI | Pallene | above, p. 182 |
| O | 182/1 | Timesianax | XII | Probalinthos | *Hesperia*, XXVI, 1957, pp. 66, 94 |
| I* | 181/0 | Hippias | I | Lamptrai | above, pp. 195-200 |
| O | 180/79 | | 2 | | above, p. 199 note 11 |
| O* | 179/8 | Menedemos | III | Prasiai | |
| O* | 178/7 | Philon | IV | Potamos | above, pp. 158-159 |
| I | 177/6 | [– – –i]ppos | V | Oinoe | above, pp. 200-201; *Hesperia*, XVI, 1947, p. 188; XXVI, 1957, p. 38 note 28 |
| O | 176/5 | Hippakos | VI | Iphistiadai | above, pp. 144-145; *Hesperia*, XXVI, 1957, pp. 69-71 |
| I* | 175/4 | Sonikos | VII | Perithoidai | *Hesperia*, XXVI, 1957, pp. 68-69 |
| O | 174/3 | Alexandros | VIII | Pithos | *Hesperia*, XXVI, 1957, pp. 71-72 |
| I* | 173/2 | Alexis | (not inscribed) | | above, p. 159; *Hesperia*, XXVI, 1957, pp. 33-38 |
| O | 172/1 | Sosigenes | 10 | | |
| I* | 171/0 | Antigenes | XI | Alopeke | above, pp. 160-161, 182, 184 |
| I* | 170/69 | Aphrodisios | 12 | | above, pp. 181-182, 184, 198-199; *Hesperia*, XXVI, 1957, p. 38 note 28 |
| O* | 169/8 | Eunikos | I | Kephisia | above, pp. 143-144 |
| O | 168/7 | Xenokles | II | Teithras | above, p. 182 |
| I* | 167/6 | Nikosthenes | 3 | | above, pp. 182-183 |
| O* | 166/5 | Achaios | IV | Eupyridai | above, pp. 175-176, 183-184 |
| O* | 165/4 | Pelops | V | Hekale | above, p. 184 |
| O* | 164/3 | Euergetes [36] | VI | Kephale | above, pp. 164-165, 184 |
| O* | 163/2 | Erastos | VII | Epikephisia | above, p. 184 |
| I* | 162/1 | Poseidonios | 8 | | above, pp. 181-184 |
| O* | 161/0 | Aristolas | IX | Eleusis | |
| I* | 160/59 | Tychandros | X | Marathon | above, pp. 162-163 |
| | 159/8 | Aristaichmos | 11 | | above, p. 184 |
| | 158/7 | Pyrrhos | 12 | | above, p. 184 |
| I* | 157/6 | Anthesterios | 1 | | above, p. 184 |

[36] The name of Charias as archon in 164/3 depended upon an erroneous reading in *Hesperia*, III, 1934, pp. 27-31, no. 20, line 9. See now the correction in *Hesperia*, XXVI, 1957, p. 73. In *Hesperia*, XV, 1946, pp. 221-222, no. 49, the reading ἐπὶ Χαρί[ου (?) ἄρχοντος] is no longer tenable. A date so early as 196/5 seemed to me fifteen years ago unlikely, but I think none the less that the restoration ἐπὶ Χαρι[κλέους ἄρχοντος], there being no alternative, must be correct.

# The Sequence of Years 237

| Type | Year | Archon | | Deme of Secretary | Reference |
|------|------|--------|---|-------------------|-----------|
|  | 156/5 | Kallistratos | 2 | | |
| O* | 155/4 | Mnesitheos | III | Paiania | above, p. 184 |
| I* | 154/3 | Phaidrias | 4 | | above, pp. 181, 188 |
|  | 153/2 | Speusippos | V | Phlya | Hesperia, XXVI, 1957, p. 72 with note 40, p. 95 |
|  | 152/1 | Lysiades | 6 | | above, pp. 187-188 |
|  | 151/0 | Epainetos | 7 | | above, p. 188 |
|  | 150/49 | Aristophantos (?) | 8 | | above, p. 188 |
|  | 149/8 | Zaleukos (?) | 9 | | above, p. 188 |
|  | 148/7 | Mikion (?) | 10 | | above, p. 188 |
|  | 147/6 | Archon | 11 | | |
| O* | 146/5 | Epikrates | XII | (vice VIII) Sypalettos | Hesperia, XXVI, 1957, pp. 47, 96; above, pp. 227-230 |
| O* | 145/4 | Metrophanes | I | Lamptrai | |
|  | 144/3 | Andreas (?) | 2 | | |
|  | 143/2 | Theaitetos | 3 | | |
|  | 142/1 | Aristophon | 4 | | |
|  | 141/0 | Pleistainos | V | Boutadai | |
|  | 140/39 | Hagnotheos | VI | Thorikos | |
|  | 139/8 | Diokles | VII | Thria | Hesperia, XXVI, 1957, pp. 28, 96; XXIX, 1960, pp. 76-77 |
|  | 138/7 | Timarchos | 8 | | Ath. Mitt., LXVI, 1941, p. 228; Hesperia, XXVI, 1957, p. 96 |
| I* | 137/6 | Herakleitos | IX | Anakaia | above, pp. 181, 188-190 |
| O | 136/5 | Timarchides | 10 | | |
| I* | 135/4 | Dionysios | XI | Amphitrope | above, p. 190; Hesperia, XXI, 1952, p. 360 |
| I* | 134/3 | Nikomachos | 12 | | above, pp. 181, 190 |
|  | 133/2 | Xenon | 1 | | |
|  | 132/1 | Ergokles | 2 | | |
| O* | 131/0 | Epikles | III | Angele | |
|  | 130/29 | Demostratos | 4 | | above, p. 187 |
|  | 129/8 | Lykiskos | V | Berenikidai | Ath. Mitt., LXVI, 1941, pp. 181-195 |
| O* | 128/7 | Dionysios | VI | Kephale | Hesperia, XXIV, 1955, p. 230 |
| I* | 127/6 | Theodorides | VII | Thria | Hesperia, XXVI, 1957, pp. 228-229; Pritchett and Neugebauer, Calendars, p. 77 |
| O | 126/5 | Diotimos | 8 | | |
| I* | 125/4 | Jason | IX | Eleusis | above, pp. 181, 190-191 |
|  | 124/3 | Nikias & Isigenes | X | Phaleron | |
|  | 123/2 | Demetrios (?) | 11 | | |
| O* | 122/1 | Nikodemos | XII | Oinoe | |
|  | 121/0 | Phokion (?) | 1 | | |
|  | 120/19 | Eumachos | II | Diomeia | |
| I* | 119/8 | Hipparchos | III | Paiania | Kerameikos, III, pp. 2-3; above, pp. 181, 191 |
| O* | 118/7 | Lenaios | IV | Skambonidai | |

| Type | Year | Archon | Deme of Secretary | | Reference |
|------|------|--------|--------|---|-----------|
| O | 117/6 | Menoites | 5 | | |
| I* | 116/5 | Sarapion | VI | Iphistiadai | *Hesperia*, XXVI, 1957, p. 97 |
| | 115/4 | Nausias | 7 | | |
| | 114/3 | [--]ratou (gen.) | 8 | | |
| I* | 113/2 | Paramonos | 9 | | above, pp. 181-191 |
| O* | 112/1 | Dionysios | X | Rhamnous | |
| | 111/0 | Sosikrates | XI | Krioa | |
| | 110/09 | Polykleitos | 12 | | above, p. 4; *B.C.H.*, LIX, 1935, pp. 66-67 |
| O* | 109/8 | Jason | I | Lamptrai | *B.C.H.*, LIX, 1935, pp. 66-67 |
| I | 108/7 | Demochares | II | Ankyle | above, p. 187 with note 26 |
| O* | 107/6 | Aristarchos | III | Paiania | |
| O* | 106/5 | Agathokles | IV | Aithalidai | |
| I | 105/4 | | 5 | | *Hesperia*, XXVI, 1957, p. 28 |
| O* | 104/3 | Herakleides | VI | Hermos | *Hesperia*, XXVI, 1957, pp. 25-28 |
| O* | 103/2 | Theokles | VII | Kothokidai | *H.S.C.P.*, LI, 1940, p. 117 |
| I | 102/1 | Echekrates | 8 | | *H.S.C.P.*, LI, 1940, p. 118 |
| O* | 101/0 | Medeios | IX | Eleusis | *H.S.C.P.*, LI, 1940, p. 119 |
| | 100/99 | Theodosios | 10 | | *H.S.C.P.*, LI, 1940, p. 120 |
| | 99/8 | Prokles | 11 | | *H.S.C.P.*, LI, 1940, pp. 121, 123 |
| | 98/7 | Argeios | 12 | | Daux, *R.E.G.*, XLVII, 1934, pp. 164-179; Dow, *H.S.C.P.*, LI, 1940, pp. 121, 122 |
| | 97/6 | Herakleitos | 1 | | *H. S. C. P.*, LI, 1940, p. 123 |
| | 96/5 | [--]kratou (gen.) | 2 | | Dinsmoor, *Athenian Archon List*, p. 202 |
| O* | 95/4 | Theodotos | III | Paiania | *Hesperia*, XVII, 1948, pp. 25-26 |
| | 94/3 | Kallias | 4 | | Dinsmoor, *Athenian Archon List*, p. 203 |
| | 93/2 | Kriton | 5 | | *ibidem* |
| | 92/1 | Menedemos | | | Dinsmoor, *Athenian Archon List*, p. 204 |
| | 91/0 | Medeios | | | *I.G.*, II², 1713 |
| | 90/89 | Medeios | | | *ibidem* |
| | 89/8 | Medeios | | | *ibidem* |
| | 88/7 | ἀναρχία | | | *ibidem* |
| | 87/6 | Philanthes | | | *I.G.*, II², 1713; *Hesperia*, Suppl. VIII (1949), p. 117 |
| | 86/5 | Hierophantes | | | *ibidem* |
| | 85/4 | Pythokritos | | | *ibidem* |
| | 84/3 | Niketes | | | *ibidem* |
| | 83/2 | Pammenes | | | *Hesperia*, Suppl. VIII (1949), p. 117 |
| | 82/1 | Demetrios | | | *ibidem* |
| | 81/0 | Ar[---) | | | *ibidem* |

# CHAPTER XII

## Conclusion

Many problems about the calendar remain unsolved. One can only hope that new evidence will throw light on some of the darker spots. This new evidence will probably be largely epigraphical, and to some extent numismatic. We owe to study in these fields our better knowledge even now of the calendar of the Hellenistic age, though the evidence of the coins can as yet be used only tentatively and awaits confirmation from the further analysis of Athenian New Style tetradrachms now in progress.

Nothing has been said in the foregoing pages about the date of the battle of Plataia, or of the battle of Marathon.[1] The latter, at least, is a problem of textual and historical interpretation as much as a problem of the calendar. About the former, Plutarch's account shows a discrepancy between dates at Athens and Plataia which argues irregularity with respect to the moon in Plataia rather than in Athens. The known examples of tampering with the calendar in Athens served always to retard, not to advance, the Athenian date. When Aristoxenos cites a hypothetical date as the tenth of the month at Corinth, the fifth at Athens, and the eighth somewhere else,[2] the irregularity may indeed have been attributed to Athens (as the retarded date) but since no definite occasion is mentioned this becomes merely evidence that all calendars did not always agree. The evidence of Aristophanes in the *Clouds* and in the *Peace* for calendar irregularity must be considered in connection with the dates for 423 and 421 B.C. as given by Thucydides. Both dates fell in

---

[1] See Pritchett, *B.C.H.*, LXXXI, 1957, pp. 277, 278-279.
[2] *Harmonica*, II, 37; cf. Pritchett, *B.C.H.*, LXXXI, 1957, p. 276.

Elaphebolion and came at that time of year when irregularities
in Athens (when such existed) were especially apt to occur.[3]
It would be unwise to press the evidence of comedy too much in
detail, but one should note that the gods came expecting their
feast at an appointed time (no doubt κατὰ θεόν) and had to go
home without.[4] Apparently the date κατ' ἄρχοντα was delayed,
though this interpretation is not inevitable. Yet it agrees with
the later evidence about the effect which archon's tampering
always had upon the calendar.

For the rest, we must keep in mind the whole body of the
evidence, not merely the exotic and non-conformist detail, and
hold fast to the conclusions reached above: (1) that the first day
of the month was determined not by observation of the crescent
moon but by a rule of convenience, (2) that this rule of convenience was that of alternating full and hollow months, probably checked from time to time with reference to the moon,
(3) that the rules of Aristotle and of Pollux about lengths and
sequences of prytanies should be applied with a measure of
flexibility, and (4) that this same flexibility existed also in the
fifth century, where the conciliar year approximated the seasonal
year.

It is inevitable that most of the time our knowledge of this
calendar must lack precision. The translation of dates in the
fifth century into Julian time must be only approximate, usually
"within a few days," though we are on firmer ground in the
summer of 422 B.C., which offers one of the two available equations allowing comparison of the conciliar year with the festival
year, and hence with the moon and with Julian reckoning.

We are again on firm ground in the late fourth century.[5] The

---

[3] See above, pp. 161-166. Pritchett, *B.C.H.*, LXXXI, 1957, pp. 277-278, 279, has
noted these dates and the comments of Aristophanes as evidence for the
archon's tampering with the calendar at Athens.
[4] *Clouds*, lines 615-619.
[5] See above, pp. 132-134.

evidence is so nearly complete, our knowledge of the sequence of ordinary and intercalary years so good, and our knowledge of the sequences of months and prytanies within each individual year so reasonably good (even to the irregularities), that dates either by prytany or by month can be translated into the Julian calendar with confidence: correct for all practical purposes, usually within a day.

Later on, where considerable sequences of ordinary and inter-calary years are known, one can usually be confident also of Julian dates. But there is always danger of being just one Attic month in error if the time of an intercalation is not certainly known. Additional errors, measured in days, may occur if there has been tampering with the calendar on the part of the archon.

Our investigation has shown that this tampering was not so prevalent as has sometimes been assumed, that evidence of it (when it did occur) was usually present in the very designation of the tampered date in question, and that there were certain times within the year, especially in Elaphebolion, when dates κατ' ἄρχοντα were more than usually liable to retardation behind the dates κατὰ θεόν.

There always remains, of course, the chance that a date κατ' ἄρχοντα may differ from its corresponding date κατὰ θεόν without our being aware of it. So far as the evidence from the fourth to the first century can show, the chance is very slight. I suggest that for those periods where we lack precise knowledge, and where the order of intercalations is uncertain, the best working rule is to take every date as a true date within its month, and every month as a normal month within its year. This means that, normally, we assume the first day of the festival year to have fallen on the first day after the summer solstice when the crescent of the new moon was visible. Paradoxically, though the Athenian calendar was not the astronomical calendar, and though the observations of the astronomers have no value in fixing dates in the actual calendar of Athens, it is in fact the astronomical calendar that can thus best be used to fill in, hypothetically, those gaps where we do not know from better evidence what the calendar at Athens really was.

A further word of caution should be added. In this volume the first day of the festival month has been taken to be, normally, that day on which the new crescent was first visible. But we have noted that the Athenians did not determine the beginnings of their months by continuous observation. They had a rule of convenience and they may have scaled this to come closer to what they thought to be the time of true conjunction, with correction from time to time by observation of the crescent. We cannot, I think, arrive at precision, and in any given reckoning we have to make allowance for this additional slight margin of error.

# INSCRIPTIONS STUDIED OR EMENDED *

## Agora Inventory Numbers

## American Journal of Philology

## Carl Curtius, Inschriften und Studien zur Geschichte von Samos

## W. B. Dinsmoor, Archons of Athens

## Epigraphical Museum Inventory Numbers

## Hesperia

* An asterisk signifies a difference of reading or restoration.

## Inscriptiones Graecae

### Inscriptions de Délos

### Pritchett and Meritt, Chronology

### Pritchett and Neugebauer, Calendars

Traversa, August, 224 with note 9, 225

Travlos, John, 192 note 1

Tréheux, Jacques, 184 with note 21

Twentieth day of a month, designations for, 45, 46 note 6, 58-59, 59 note 37, 146 note 33

Twenty-first day of a month, designations for, 45, 46 note 6, 47 note 7, 58-59, 104 note 48

Tybi, Egyptian month, 20, 21, 22, 28

Tychandros, archon in 160/59, 162-163, 236

Unger, Georg F., 92, 178 note 20

Usener, Hermann Karl, 93

van Leeuwen, Jan, 43, 45 note 9

Venetus manuscript of Aristophanes, 43-44, 45 with note 9, 46 note 4

Wade-Gery, H. T., 4 note 2, 46 note 1, 75 note 9, 93 note 29, 216 note 30, 218 note 36

White, John Williams, 43 with note 6

Wilhelm, Adolf, 122

Woodhead, A. G., 214 note 24, 232 note 34

Woodward, Arthur M., 68 note 19

Wycherley, Richard E., 192 note 1

Xanthos, father of Ameinias of Marathon, 200

Xenokles, archon in 168/7, 182, 236

Xenon, archon in 133/2, 237

Xenophon, archon in 287/6, 232

Year, analysis of in 407/6, 212-214; analysis of in 336/5, 14-15; analysis of in 307/6, 176-178; analysis of in 222/1 and 221/0, 172-175. See also Conciliar year, Festival year, Intercalary year, Ordinary year, Seasonal year, Sidereal year, Solar year

Years, numismatic evidence for sequence of, 180-191; sequence of (346/5-81/0), 231-238

Zacher, Konrad, 43 note 6

Zaleukos (?), archon in 149/8, 188, 237

Zeno, age of, 224; death of, 221

Zopyros, archon in 186/5, 236